PRINCIPLES OF PLASTICS EXTRUSION

Principles of PLASTICS EXTRUSION

J. A. BRYDSON
National College of Rubber Technology, Polytechnic of North London

D. G. PEACOCK
Bakelite Xylonite Ltd, Manningtree, Essex

APPLIED SCIENCE PUBLISHERS LTD
LONDON

APPLIED SCIENCE PUBLISHERS LTD
RIPPLE ROAD, BARKING, ESSEX, ENGLAND

ISBN: 0 85334 563 5

WITH 46 ILLUSTRATIONS

Printed in Great Britain by Galliard (Printers) Ltd Great Yarmouth

CONTENTS

INTRODUCTION

AT some time or other the bulk of thermoplastics materials produced are subjected to an extrusion operation. This is because extrusion is not only used for making end products but also for such intermediate processes as compounding, blending, pelletising, granulating, straining and preparing blanks. The extruder barrel is also incorporated in effect in much injection moulding and blow moulding equipment.

This programme teaches the basic elements of extrusion. It has been designed for readers already educated to GCE 'O' level standard who come within the following categories:

1. Students or industrial trainees who:
   (a) will specialise in extrusion. The programme will give them an early and precisely designed foundation;
   (b) will be required to understand the process as a part of a general training in plastics processing;
   (c) will be involved in the process on a practical basis but will not have had any previous formal instruction in the process.
2. Company employees (for example, those engaged in sales and administration) who, because of the nature of their work, will be concerned with extrusion without necessarily being directly involved in it. Such personnel could be expected to benefit in terms of improved understanding and communication.
3. Pupils in schools (for example, science sixth-formers) who already have an interest in the plastics industry and wish to learn more about one of the processes in this industry.

HOW TO WORK THROUGH THE PROGRAMME

This teaching programme consists of a series of short paragraphs called *frames*. In most frames you will be asked a simple question about what you have been learning. You may be asked to select an answer from some alternatives or to fill in missing words.

Read each frame carefully, write down your answer and then check it with the correct answer which is printed beneath the band. With each programme a mask should be used to cover up the correct answer until you have written down your own answer. Do not forget to put the mask in position each time you turn to a new page. In some questions missing words are indicated by one or more straight lines, the number of which indicates the number of words in the answer. Where the blank shows a row of dots this means that the answer consists of a phrase or sentence. Do not write answers in these spaces, write them on a separate sheet of paper.

HAVING worked through the programme at his own speed—the learner will be able to:

(a) describe the principles of extrusion and be aware of its advantages and limitations;

(b) recognise any machine capable of operating this process;

(c) identify the principal parts of an extruder;

(d) state what the principal parts do and the sequence in which they operate;

(e) give technical reasons why specified plastics articles can or cannot be manufactured by this process.

Acknowledgement

Several of the diagrams are based on the Plastics Institute Monograph on Extrusion by E. G. Fisher and we are grateful to be able to use these in this book.

PART ONE

THE NATURE OF PLASTICS

Frame 1.
It is not easy to give an exact definition of plastics. Although some natural plastics like shellac and bituminous plastics exist, most are prepared by methods of chemical synthesis. These may be described as ________ materials.

synthetic

Frame 2.
Most plastics, including all the commercially available thermoplastics, are ORGANIC in nature.
Which of the following are organic?
(a) silica; (b) butane; (c) sugar; (d) benzene; (e) lime; (f) natural rubber.

(b) butane; (c) sugar; (d) benzene; (f) natural rubber
If you were correct, go on to Frame 3. If you were wrong, read Frame 2*.

Frame 2*.
The term ORGANIC is applied to those chemical compounds in which the element carbon plays an important part (carbonates such as calcium carbonates which contain no other carbon atoms are not usually considered to be organic). Until the early nineteenth century it was believed that these compounds could be produced only with the help of living organisms—hence the term organic. This concept is now known to be incorrect, but the term organic chemistry has been retained to describe the chemistry of carbon compounds.

If the main structural element in a chemical compound is carbon, then the compound is an ________ one.

organic Now proceed to Frame 3

Frame 3.

With the exception of bituminous plastics and one or two other natural materials which are not always classified as plastics, plastics are POLYMERIC in nature.

Which of the following are polymeric?

(a) calcium carbonate; (b) glucose; (c) cellulose; (d) natural rubber; (e) protein; (f) polyethylene; (g) ethyl alcohol.

(c) cellulose; (d) natural rubber; (e) protein; (f) polyethylene

If you were correct, go on to Frame 4. If you were wrong, read Frame 3*.

Frame 3*.

POLYMERS are large molecules built up by repetition of a very limited number of much smaller molecules, usually of only one or two types. Such simpler molecules are known as MONOMERS. The process of joining together such monomers to produce polymers is known as POLYMERISATION.

Polymers are produced by the polymerisation of ________.

---

monomers

Frame 3**.

Many polymers occur naturally. Proteins may be considered to be made up by the joining together of amino acids, cellulose from glucose and natural rubber from isoprene (in the case of these natural products this does not necessarily mean that the polymers have been made this way in nature but simply indicates the nature of the repeating units).

Many other polymers are prepared synthetically. For example, polyethylene is made by the ________ of ethylene.

---

polymerisation

Frame 4.
What polymers are produced from the following monomers?
(a) propylene; (b) vinyl chloride; (c) butadiene;
(d) acrylonitrile.

(a) polypropylene; (b) polyvinyl chloride (p.v.c.); (c) polybutadiene;
(d) polyacrylonitrile

Frame 5.
Polybutadiene is normally a rubber and polyacrylonitrile is a fibre. Polypropylene and p.v.c. are well-known plastics but can be produced in fibre form. Hence, although plastics are usually polymers, polymers are not necessarily plastics.

Name another polymer which is well known in both fibre and plastics form.

nylon

Frame 6.
Name three normal characteristics of plastics materials.

(a) synthetic; (b) organic; (c) polymeric

By tradition rubbers are not considered as plastics so that a fourth characteristic should be added. Plastics should be non-rubbery or non-elastomeric (at least at temperatures of use).

Natural rubber is not normally considered to be a plastics material because it is __________.

elastomeric (rubbery)

Note that many surface coatings, adhesives and fibres have the same four characteristics as plastics. The first two could be considered as plastics in two dimensions and fibres as plastics in one dimension. It is, however, normal to consider such groups as separate from plastics materials.

Frame 7.

Polymers are seldom used as plastics uncompounded (unmixed) with ADDITIVES. For example, p.v.c. is frequently mixed with whiting (chalk) as a FILLER, PLASTICISERS to improve flexibility, STABILISERS to improve the resistance to heat and light, PIGMENTS, and LUBRICANTS to improve flow properties and prevent the compound sticking to processing machinery.

Tribasic lead sulphate is often used to improve the resistance of p.v.c. to heat and light. Used for this purpose the ADDITIVE is known as a ________.

stabiliser

Frame 8.

In order to make articles of the correct shape it is normally necessary to be able to make the plastics material flow.

In what physical forms are the following common materials when we make them flow?

(a) toothpaste; (b) sealing wax; (c) polystyrene cement.

(a) a paste (suspension); (b) molten (liquid); (c) liquid (solution)

Plastics materials are most commonly processed in a viscous molten state, which allows them to be shaped and moulded by flow. The fact that plastics can be made to ____ is another important characteristic.

flow

Frame 9.
Usually plastics materials will only flow in the viscous molten state under the influence of heat and pressure. Thus moulding and extrusion processes used with plastics normally involve the application of ______ and ________.

---

heat; pressure

It should be noted that, apart from being inorganic, glasses possess most of the characteristics of plastics but like rubbers are arbitrarily considered as a different class of materials.

Frame 10.
Plastics may be conveniently divided into two groups, THERMOPLASTICS and THERMOSETTING PLASTICS (THERMOSETS).

Thermoplastics when heated to a sufficiently high temperature will soften and flow under pressure. On cooling they harden. This process may then be repeated over and over again. (Note: continual heating may lead to degradation processes which will put a limit on the number of times this process may be repeated.)

The diagram

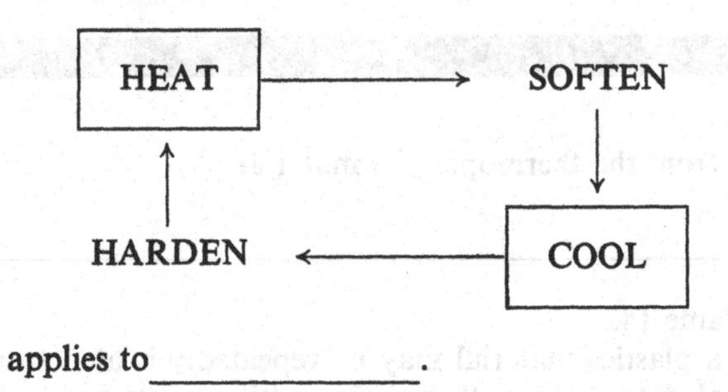

applies to ______________.

---

thermoplastics

Frame 11.
Thermosetting plastics when heated once will soften and flow under pressure. However, simultaneously, chemical reactions occur which cause the material to harden and set. Even though more heat is applied they will not ________ again.

---

soften

Frame 12.
The 'setting' is due to chemical reaction which causes the polymer molecules to cross-link one with another so as to inhibit flow.

Therefore flow is prevented because the molecules have ____________.

---

cross-linked

Frame 13.
If you have one article made from a thermoplastics material and one from a thermosetting plastics material, which one would soften on heating?

---

the one made from the thermoplastics material

Frame 14.
If a plastics material may be repeatedly heated to soften and cooled to harden then it is a (thermoplastic/thermosetting) plastic.

---

thermoplastic

Frame 15.
Since the extrusion of most commercial plastics is carried out with thermoplastics materials this programme will be restricted to discussing extrusion of these materials. The process of heating thermoplastics to soften them is known as plasticising. Therefore in order to plasticise materials in this way it is necessary to _____ them.

heat

Frame 16.
The word PLASTICISER as already noted refers to an additive used to improve flexibility. Such additives are normally non-volatile liquids which are blended with the polymer and may not only increase the flexibility but also reduce melt viscosity.

Hence ___________ p.v.c. is more flexible than unplasticised p.v.c. (UPVC).

plasticised

Frame 17.
When thermoplastics have been thermally plasticised (*i.e.* by heating) they can be made to flow under pressure.

So the two things needed to make thermoplastics flow are _____ and _________.

heat; pressure

Frame 18.

There are many means by which thermoplastics materials may be processed using heat and pressure. One important method is extrusion. In this process softened material is forced continuously through a die to give a product of constant cross section.

In order to produce a product of constant cross section by an extrusion process the material is forced continuously through a

__________.

---

die

Frame 19.

Well-known extruded products are polyethylene film, p.v.c. guttering and piping, high-impact polystyrene sheet, p.v.c. insulated flex, polymethyl methacrylate (acrylic) light fittings and nylon monofilament.

Name five thermoplastics materials which may be extruded.

---

polyethylene; p.v.c.; high-impact polystyrene; polymethyl methacrylate; nylon

---

Frame 20.

A simple example of everyday extrusion is seen when toothpaste is squeezed out of the tube. Providing the material squeezed out (the extrudate) is not deformed by stretching or gravity or other influence, it has a constant cross section.

This process is limited and said to be discontinuous because the length of extrudate produced is limited to the capacity of the tube.

Toothpaste extrusion is therefore a __________ extrusion process.

---

discontinuous

Frame 21.
In piping of icing onto cakes a simple ram extruder is used. A hand-operated plunger or ram forces icing out of a cylinder through a small orifice (die). When the barrel is empty it may be possible to withdraw the plunger, feed in more icing and again start depressing the plunger without breaking the thread of icing.

Ram extrusion can therefore be considered as a semi-continuous process.

Which of the following two processes may be considered as semi-continuous?

(a) squeezing of toothpaste out of a tube;
(b) piping icing onto a cake.

(b) piping icing onto a cake

Frame 22.
For extrusion of thermoplastics it is very desirable to be able to force material continuously through a die. This can be effected by roller pumps, gear pumps or screw pumps, the latter being the most commonly used.

Screw pump extrusion is generally preferred to ram extrusion since it is a ____________ extrusion process.

---

continuous

Frame 23.
It is usual to feed material to a screw pump *via* a hopper. A typical thermoplastics extruder therefore may be considered as consisting of three parts. What are they?

---

a hopper, a screw pump and a die

Frame 24.
Figure 1 shows a schematic extruder with parts labelled A, B and C. Which of these is the hopper, which is the pump and which is the die?

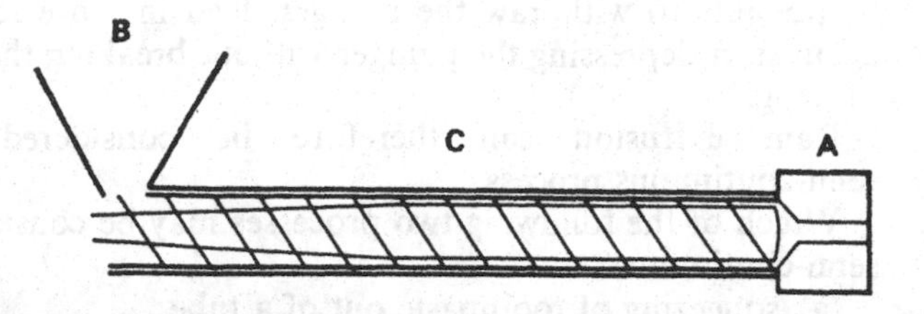

Fig. 1

A—the die
B—the hopper
C—the screw pump

Frame 25.
In order to produce satisfactory extrudates it is necessary to apply heat to the granules in order to soften them and make the resulting melt capable of flow. This is normally carried out in the screw pump part of the extruder so that the screw pump has the additional function of being a heating chamber.

Thus, so far we have seen that the screw pump has two functions in an extruder. These are to ______, ______ the thermoplastic and ______ it through the die.

heat, soften (melt); pump

Frame 26.
The heat required to soften the granules is obtained by the use of external heaters surrounding the pump and also by frictional heat generated while working the thermoplastics material.

In addition to heating the granules with external heaters, heat may also be obtained by __________.

friction (frictional working)

Frame 27.
In order to ensure a smooth extrudate it is necessary to mix the melt so that it has uniform flow properties and to compress it. There are therefore two additional roles of the screw pump section of the extruder, *i.e.* it is a mixing device and a pressure vessel.

What are the four roles of the screw pump?

pump; heating chamber; mixing device; pressure vessel

Since this part therefore has so many functions the term screw pump is commonly replaced by the term 'barrel and screw'.

It is convenient to consider the extrusion process in three sections:

(a) barrel and screw;
(b) the die;
(c) the haul off and ancillary equipment.

The rest of this programme will deal in turn with these three sections but before going on to the first of these see if you can answer the questions in Criterion Test Part 1.

CRITERION TEST

Part One

1. Name four characteristics of plastics which help to distinguish them from other materials.

2. Can you name any other classes of materials that can have these characteristics?

3. The two main groups of plastics are ____________ plastics and ____________ plastics.

4. What is the difference in behaviour observed when heating finished articles made from these materials?

5. _____ and _________ are normally applied to a thermoplastics material in order to make it flow.

6. Name four functions of the barrel and screw (screw pump) section of an extruder.

PART TWO

THE BASIC SINGLE-SCREW EXTRUDER

Frame 1.

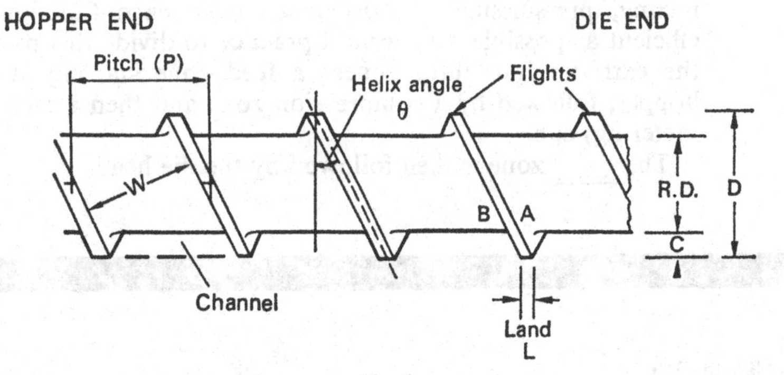

Fig. 2

Figure 2 shows typical screw nomenclature

P	is the pitch
D	is the screw diameter
W	is the channel width
C	is the channel depth
L	is the land width
R.D.	is the root diameter
θ	is the helix angle
A	is the screw leading edge
B	is the screw trailing edge

Are the following statements true or false?

(a) the difference between the screw diameter and the root diameter is the channel depth;

(b) the helix angle is the angle between the screw thread and the transverse plane of the screw;

(c) the projection of channel width plus land width on to the screw axis is equal to the screw pitch;

(d) the helix angle is equal to the root diameter divided by the land width.

(a) false (D − R.D. = 2C); (b) true; (c) true; (d) false

Frame 2.

It was seen in Part One that the barrel and screw section of the extruder had four principal functions (pumping, heating, mixing, pressurising). In order to make each function as efficient as possible it is normal practice to divide this part of the extruder into three zones; a feed zone starting at the hopper, followed by a compression zone and then a melt (or metering) zone.

The ______ zone is then followed by the die head.

melt (metering)

Frame 3.

The function of the feed zone is to collect granules from the feed hopper and transport (pump) them up the screw channel. At the same time the granules should begin to heat up and compact, building up pressure as they advance down the screw. For efficient pumping the granules must not be allowed to lie in the screw channel. They must therefore show a high degree of slippage on the screw channel surface and a ______ degree of slippage on the barrel wall.

low

Frame 4.

Maximum delivery of granules by the feed section may be achieved by:

(a) a relatively deep channel (in comparison with the rest of the screw);
(b) a low degree of friction between the granules and screw;
(c) a high degree of friction between the granules and barrel wall;
(d) an optimum helix angle (usually about 20° for polyethylene)—many screws are designed with the pitch equal to the diameter but in some cases it may be slightly less.

With many polymers, such as polyethylene, it is found that the friction of polymer to metal increases with temperature up to about 120°C. For optimum pumping we should therefore in theory try to have a hot/cold screw in this zone and a hot/cold barrel.

cold; hot

In practice it is invariably found that screw cooling reduces output. This is probably due to other effects occurring further along the screw.

Frame 5.

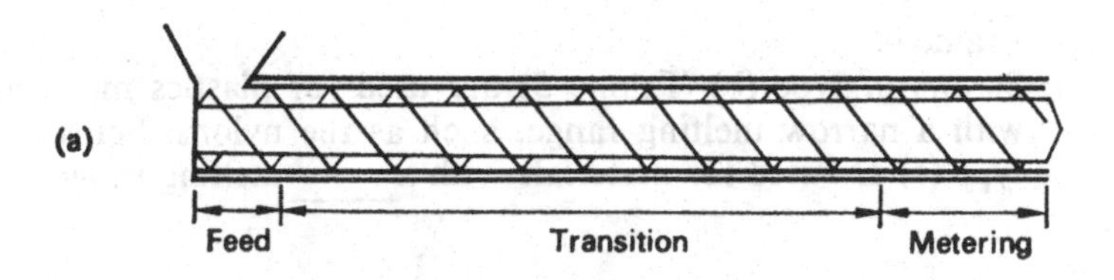

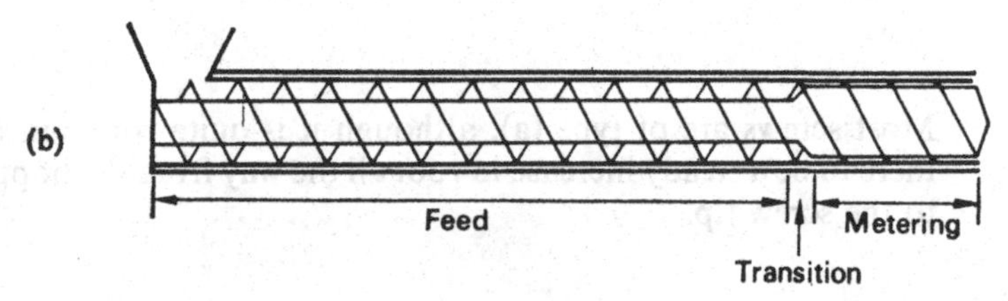

Fig. 3

In screw (b) the compression (transition) zone is much shorter than in screw (a), but in each case it will be seen that as the screw goes from the feed zone to the melt zone there is an increase in the screw root. At the same time there is a decrease in the volume of the space enclosed by the thread and the surface of the root in one complete turn of the screw. This will cause compression of the granules forcing air between the granules back towards the hopper. Granule melting should occur around the compression zone in order to consolidate the polymer.

Can you suggest an alternative way by which a screw with constant screw diameter can incorporate a compression zone?

---

use a screw with a decreasing pitch (see diagram below)

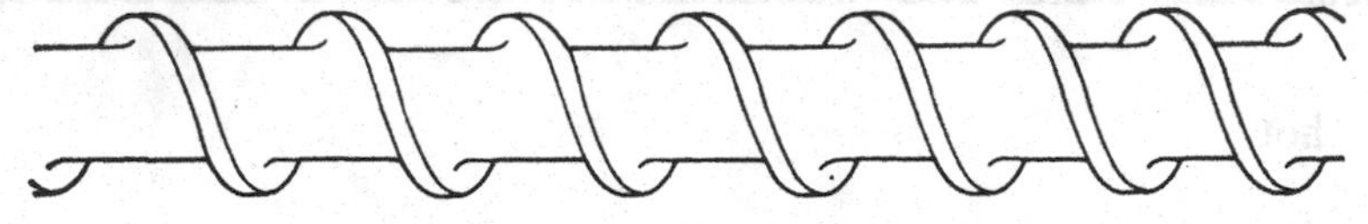

Extruder screw with diminishing pitch and constant root diameter

Another possibility, but not known to have ever been used in practice, is a screw with decreasing channel width and increasing land width. Decreasing pitch screws are useful where it is desired to reduce the amount of shear on the polymer melt.

Frame 6.
Screws of type (b) (Frame 5) are used for plastics materials with a narrow melting range, such as the nylons. Screws of type (a) are used for materials with a _____ melting range.

---

wide

Most screws are of type (a), although it is quite common for there to be a steady increase in root all the way from the hopper to the screw tip.

Frame 7.

The ratio of the volume of the first turn of the channel of the screw (at the hopper end) to the volume of the last turn of the channel (at the delivery end) is known as the compression ratio. In commercial machines this usually has a value between about 1·5:1 and 4:1.

Which of the two screws shown in Fig. 4 has the higher compression ratio?

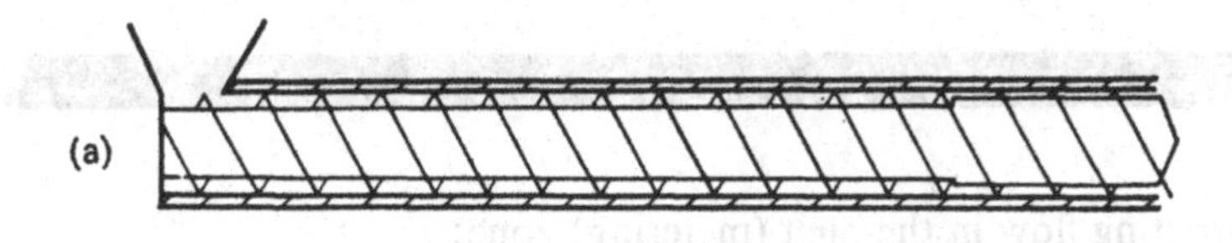

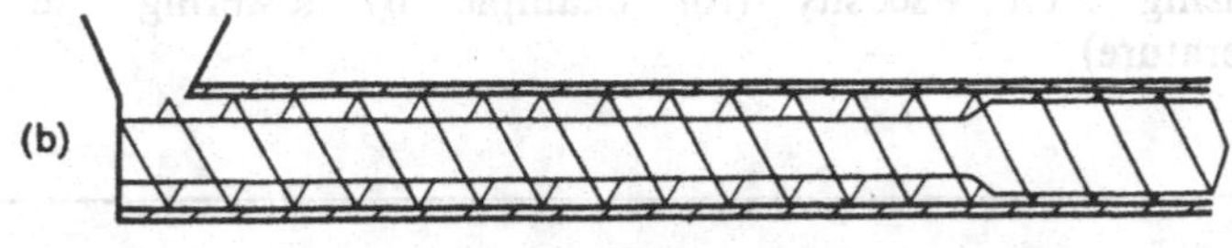

Fig. 4

screw (b)

Frame 8.

In the melt (metering) zone the polymer melt is brought to the correct consistency and pressure required for extrusion. The melt should be pumped to the die at a constant rate, consistency and pressure.

(These three properties may vary from point to point but when measured at a particular point they should not change with time.)

The function of the melt zone is to bring the melt to the correct ____________ and ________ required for extrusion.

consistency; pressure

Frame 9.
A high melt pressure is required in the metering zone in order to mix the melt to give it constant properties throughout and hence obtain smooth extrudates.

This pressure is generated by restrictions to flow in the melt zone and in the die head. It will also increase with an increase in melt viscosity.

Give three ways of generating a pressure in the melt.

(a) restricting flow in the melt (metering) zone;
(b) restricting flow in the die head;
(c) increasing melt viscosity (for example by lowering the melt temperature)

Frame 10.
Restrictions to flow in the metering zone may be increased by:
(a) decreasing channel depth;
(b) decreasing channel width;
(c) replacing all or part of the metering zone section of the screw with a smear head attached to and revolving with the screw (see diagram);

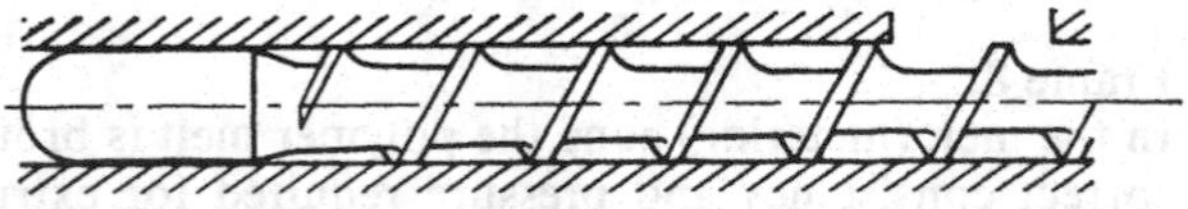

(d) water cooling of the screw.

Which of the above methods may normally be used without purchasing a new screw?

(d) screw cooling facilities are usually supplied. (Note: this is a second function of screw cooling—*see* Frame 4.)

Frame 11.
Screw cooling is effective in that it chills layers of polymer adjacent to the root surface and reduces the effective channel depth.

A smear head also has the same effect in increasing the restriction. It can have additional advantages in increasing frictional heating (when desirable), increasing the degree of mixing of the melt and damping out any pulsations in output.

What function does a smear head have in common with a cooled screw?

---

increasing the melt pressure (in the melt zone)

Frame 12.
At the end of the melt zone there is often a breaker plate (*see* Fig. 5).

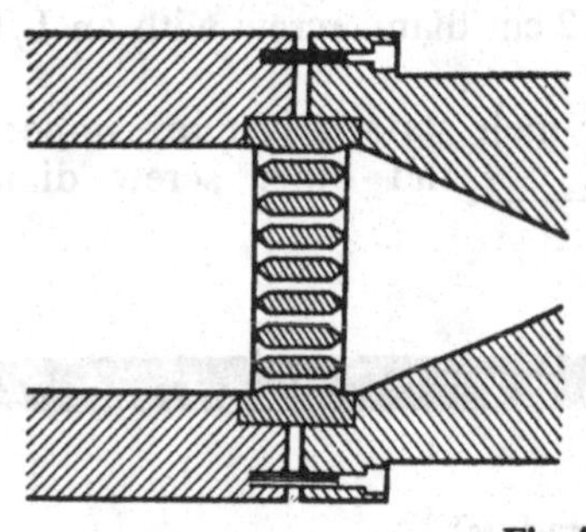

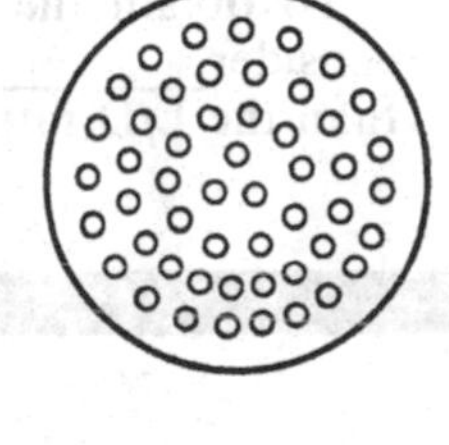

Fig. 5

This is usually a perforated disc of slightly greater thickness than the sum of two recesses cut in the barrel and the die head (or die adaptor).

It has several important functions:

(a) it helps to further increase back pressure;

(b) it turns rotational flow of the melt into flow parallel to the screw axis;

(c) it holds back impurities;

(d) it holds back unplasticised material.

The 'power' of a screw pack to fulfil these functions may be increased by interposing wire mesh screens between breaker plate and screw.

If the thickness of breaker plate plus screen pack were smaller than the sum of the depth of the two recesses, polymer melt would collect in these spaces, stagnate and_________.

decompose (degrade)

Frame 13.
Screws should have a sufficient length and diameter in order to accommodate the feed, compression and metering zones so that the melt is in the correct state when it enters the die. Screw dimensions such as helix angle, channel depth and width are also significant here. The common practice of using the length/diameter ratio to assess the ease with which the melt state requirement can be met is therefore an over-simplification. For example, a 15 cm diam. screw with a length/diameter ratio of 10/1 would be expected to bring the melt into the correct state more easily than a 2 cm diam. screw with an L/D ratio of 20/1.

To obtain the correct melt state it is more important to consider________, __________and other screw dimensions than the L/D ratio.

(screw) length; (screw) diameter (or radius)

In spite of the above reservation it should be noted that most thermoplastics extruders have L/D ratios of between 15/1 and 25/1.

Frame 14.
Some extruders have two screw threads. They are known as two-start or twin-start screws and are more expensive to produce. Additional disadvantages are that they may deliver at different flow rates giving a pulsation effect and also that polymer melt in one channel may become blocked without the operator's knowledge. The melt may hence become overheated and ________.

decompose (degrade)

In general, therefore, single-start screws are recommended.

Frame 15.
In order to melt the granules, heat is generated either internally by friction or by applying external heat from heaters wrapped around the barrel. It is necessary to control the heat supply because if the material becomes too hot it may decompose, degrade or become too fluid. If too cold it will be insufficiently plasticised. Variations in temperature will also cause variations in flow rate.

To prevent decomposition or degradation the melt must not be allowed to become too ____.

hot

Frame 16.
In some cases the contents of the barrel may overheat. Rapid cooling may be achieved by having channels cut in the barrel through which cooling water or forced draught air may be circulated.

Air or water may be circulated in channels round the barrel in order to extract ______ from the system.

heat

Frame 17.

In order to optimise the efficiency of the barrel and screw in its various functions it is normal practice to be able to vary the temperature along the barrel. From two to six heating zones are normal with a steady temperature gradient increasing from the feed to the die.

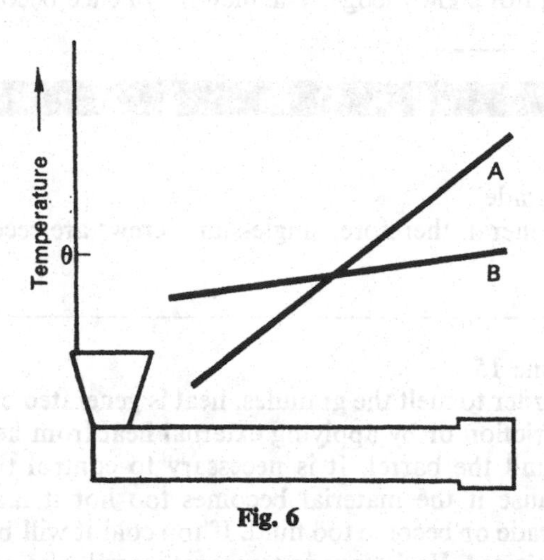

Fig. 6

In Fig. 6 are shown two temperature profiles in which similar amounts of heat are supplied to the plastics material.

Which profile do you think will be more suitable with materials which become sensitive to degradation at temperatures above, say, θ as indicated in the figure?

profile B

Frame 18.

The output of an extruder will depend on the screw dimensions, on the die dimensions and on screw speed.

The following factors will increase output:

(i) increase of screw speed (N);
(ii) increase of screw diameter (D);
(iii) the helix angle (θ) up to a maximum of about 30°;
(iv) an increase in die orifice diameter.

Place the following extruders with screw characteristics listed in an expected order of increasing output. Assume any other characteristics and conditions to be equal (in practice this may be difficult).

(a) Screw	D = 3 cm	θ = 30°	N = 20 rev/min
(b)	D = 6 cm	θ = 30°	N = 50 rev/min
(c)	D = 1 cm	θ = 10°	N = 20 rev/min
(d)	D = 1 cm	θ = 30°	N = 20 rev/min
(e)	D = 3 cm	θ = 30°	N = 50 rev/min

(c); (d); (a); (e); (b)

Frame 19.

It is a common misconception that a decrease in melt viscosity will increase output. Output will, however, depend on the ratio of the average viscosity in the die to the average viscosity in the barrel. Lower viscosities in the die compared with those in the barrel lead to increased output.

An increase in the average melt viscosity in the die compared with the average melt viscosity in the metering zone of the barrel will cause a/an __________ in output.

decrease

Frame 20.

It is common experience that output from an extruder is influenced by the form in which the thermoplastics material is fed to the hopper. In most circumstances it is found that the nearer the granules are to being spherical of homogeneous size and of about 3 mm diam. the higher is the output.

A thermoplastics material is available in the following physical forms. Put them in what you consider is the most likely increasing order of output efficiency:

(a) cube-cut granules of 27 mm^3 volume;
(b) fine powder;
(c) near-spheres of 27 mm^3 volume;
(d) 'lace cut' granules made by cutting extruded strands of circular cross section of similar volume to above;
(e) flake obtained by disintegrating sheet in an irregular fashion.

(b); (e); (a); (d); (c)

It may be found in some cases that (e) may be less efficient than (b) and (d) less efficient than (a).

The danger of spillage of granules of type (c) on to the shop floor should be pointed out (they act like ball bearings when stood on).

Frame 21.

Some materials tend to give off volatiles or gases during the extrusion process. This causes the extrudates to bubble and exhibit porosity. This can be reduced by using a screw with a decompression zone as shown in Fig. 7.

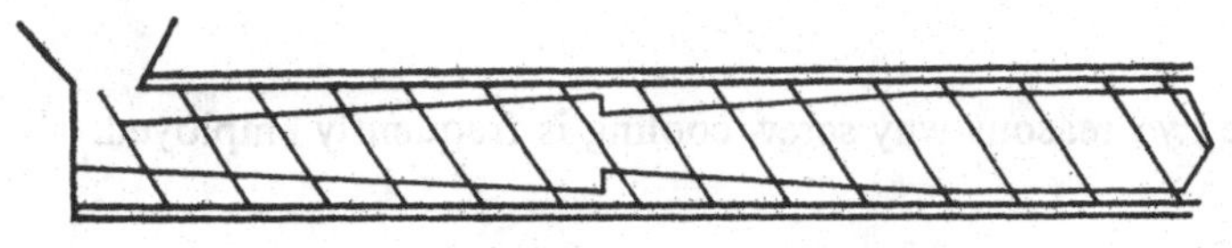

Fig. 7

Provided the back pressure is not too great the flights in the decompression zone will not be full and there will be no pressure on the melt at that point. It is thus possible to drill a vent hole through the corresponding point in the barrel to allow gases to escape—usually aided with a vacuum.

Can you suggest an alternative way of removing the volatiles when using a screw of this type?

---

drill holes through from the trailing edge of the screw flight down to a central bore to the screw; this method is less favoured since blockages would be more difficult to clean and because screw cooling would be difficult to manage.

CRITERION TEST

Part Two

1. What are the three zones of a conventional screw extruder?

2. Give two reasons why screw cooling is frequently employed.

3. Which type of screw is often preferred where it is desired to minimise the amount of shear on a polymer melt—an increasing root screw or a decreasing pitch screw?

4. Name one thermoplastics material which is usually extruded using a screw with a short compression zone.

5. Name the functions of the breaker plate.

6. Why do some screws have two compression zones?

PART THREE

PRINCIPLES OF DIE DESIGN

Frame 1.
It is the die which determines the shape of an extrudate. In this section we will consider the development of die design from those dies requiring the production of simple rods to complex sections and sheet several metres wide.

In Fig. 8 we have a typical die design for making solid rod.

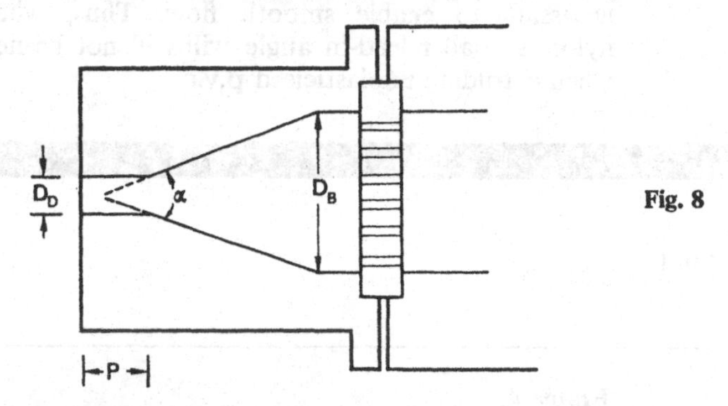

Fig. 8

On this figure the following characteristics are marked:
D_D diameter of die orifice;
D_B diameter of bore of extruder barrel;
α lead-in angle;
P die parallel or die land.

Because of the breaker plate the pressure in the melt on the barrel side of the breaker plate is greater than the pressure on the die side. At the die outlet the pressure is zero. It is, however, important to have a sufficiently high pressure in the die so that the melt is consolidated after having been strained through the breaker plate and before emerging from the die. This is achieved by designing restrictions into the die; for example, the ratio D_D/D_B must be greater/less than 1.

---

less

In practice it should normally be less than 0·5.

Frame 2.
Restriction to flow can also be increased by increasing/decreasing the length of the die parallel.

---

increasing

Frame 3.

Rods may be made from many thermoplastics including nylon, which gives a low viscosity melt under normal operating conditions, and unplasticised p.v.c., which gives a high viscosity melt.

The more viscous the melt the smaller is the lead-in angle necessary to enable smooth flow. Thus, when extruding nylon a smaller lead-in angle will/will not be necessary than when extruding unplasticised p.v.c.

will not

Frame 4.

Thermoplastic polymer molecules consist of long chains which tend to take up a randomly coiled configuration whenever possible. When such materials flow, for example when they are forced through a die, the molecules become partially straightened or oriented and are no longer randomly ______.

coiled

Frame 5.

When molten polymer emerges from a die many of its molecules will have been orientated in a direction parallel to the axis of the die orifice. When no longer constrained by the die walls the molecules tend to re-coil causing a contraction in the direction of extrusion and expansion in the cross section of the extrudate. This phenomenon is known as die swell.

For this reason extrudates, unless hauled off at a greater rate than they are extruded, show greater/lesser cross sections than those of the corresponding die orifices.

greater

Frame 6.

Under normal circumstances die swell may be reduced by the following:

(a) decreasing extrusion rate;
(b) increasing melt temperature (keeping extrusion rate constant);
(c) increasing the length of the die parallel;
(d) increasing the draw down rate (ratio of haul off rate to natural extrusion rate).

Which of the above methods would not adversely affect output, increase cooling time or involve modifications to machinery?

(d)

Frame 7.

Figure 9(a) shows a typical cross section of product obtained from a die of cross section as shown in Fig. 9(b).

What shape die (approximately) would be required to give a product of cross section as indicated in Fig. 9(b)?

(a) (b)

Fig. 9

Frame 8.

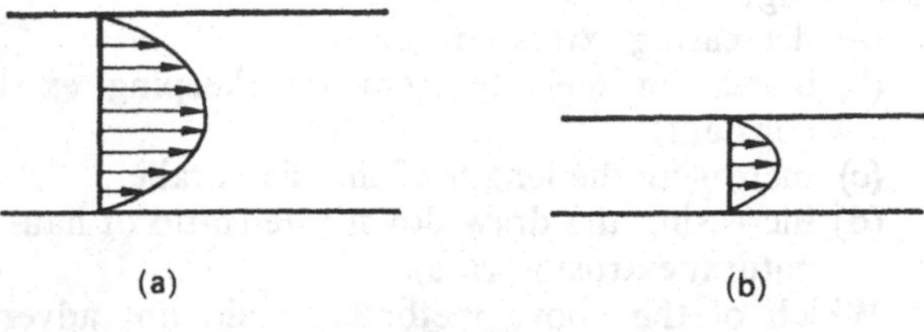

Fig. 10

Figure 10(a) shows schematically that when a polymer melt flows through a tube the rate of flow is much greater in the centre of the tube than near the walls. In fact, it is usually the case that there is no flow at the wall.

The lengths of the arrows indicate the flow velocities at various points and the volume of space which contains all the arrows (Fig. 10(a) is only a two-dimensional representation) will give an indication of the volumetric flow rate.

If we assume that the shape of the distribution of velocities is the same in Fig. 10(b) as in Fig. 10(a), which has half the radius, it is visually seen that the volumetric flow rate is less than half that in Fig. 10(a).

This indicates that for other conditions such as pressure being equal the volume flow rate is:

(a) less than proportional to the radius of the tube;

(b) proportional;

or

(c) more than proportional.

(c) more than proportional

Frame 9.
Now consider that it is required to make an extrudate of cross section as in Fig. 11.

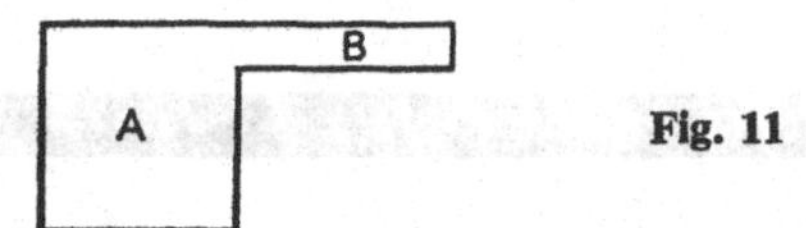

Fig. 11

Using a simple die as in Fig. 8, what do you think is likely to happen to the melt on extrusion?

the melt will flow much faster through part A than part B; the extrudate will tend to coil up on itself

---

Frame 10.
Without changing the die cross section, can you suggest a way of reducing the problem?

it is necessary to restrict the flow through part A; this may be done by having a longer die parallel at A than at B (it may also be possible to place a pin behind the A zone to act as a throttle—*see* later)

Frame 11.

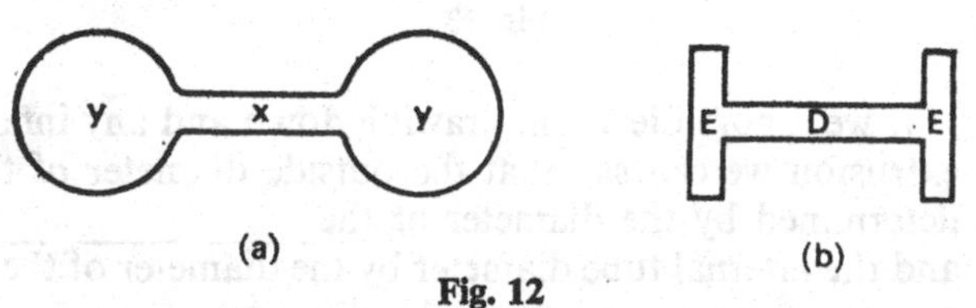

Fig. 12

It is desired to make an extrudate of cross section as in Fig. 12(a). However, the flow rate at X is much less than at parts Y.

State two results that could occur from this.

the cross section at X could be much reduced; the melt could tear periodically in a direction perpendicular to the flow

---

Frame 12.
In Fig. 12(b) the flow rate at D is greater than at E, being in the centre of the die. What do you think might happen here?

section D will show waves (ripples, undulation); alternatively there may be tearing at E

Frame 13.
So far we have only dealt with solid sections. It is often desired to make extrudates with a hollow section such as tubing.
A typical die head is shown in Fig. 13.

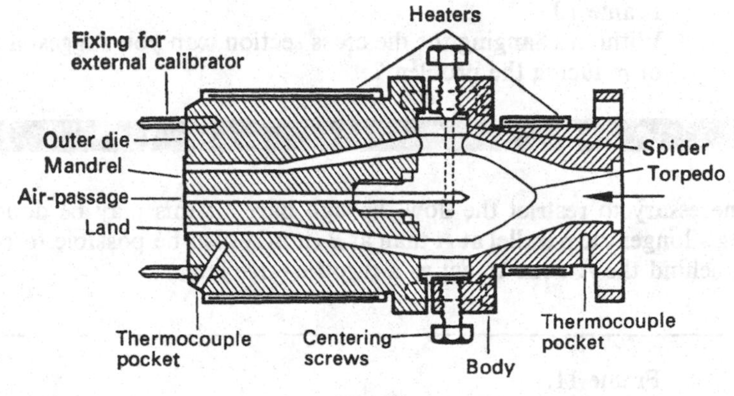

Fig. 13

If we ignore die swell, drawing down and any inflation after extrusion we can say that the outside diameter of the tube is determined by the diameter of the____ ____ ____ ____ and the internal tube diameter by the diameter of the______.

outer die ring orifice; mandrel (pin or torpedo)

Frame 14.
In order to make the mandrel and outer die ring orifice concentric, centering screws are provided to adjust the mandrel position. Unless such a facility is provided there is a danger that the product will have an ________ cross section.

eccentric

Frame 15.
The mandrel is held in position by a spider (Fig. 14). The spider legs have a cross section similar to that indicated in Fig. 14(a) to facilitate flow. It is common to drill a hole down one of the spider legs connecting an air supply to the air hole down the mandrel. This provision of compressed air allows the tube to be inflated after emerging from the die but while still molten.

In many tube extrusion processes a water-cooled sizing die is placed close to the main die and the tube inflated to the internal diameter of the sizing die. When this technique is used what factors will determine the tube internal diameter?

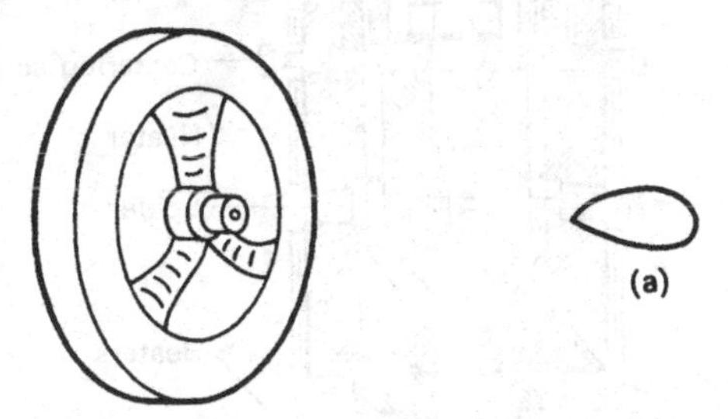

Fig. 14

(a) ratio of mandrel diameter to outer die ring orifice diameter
(b) die swell
(c) draw down rate

Frame 16.
In order to provide a smooth glossy extrudate the die head is heated usually with some form of resistance heater attached to the surface of the die head body. A cold die could also cause the polymer melt to freeze, blocking the outlet from the extrudate. Since overheating can cause such problems as degradation, control of ____________ is important.

---

temperature

Frame 17.
The type of die shown in Fig. 13 is of the in-line type. For various reasons it is sometimes desired for the extrudate to emerge from the die at an angle (most commonly at a right angle) to the axis of the extruder barrel. This may be achieved by the use of an elbow adaptor (Fig. 15) or a side entry die (Fig. 16).

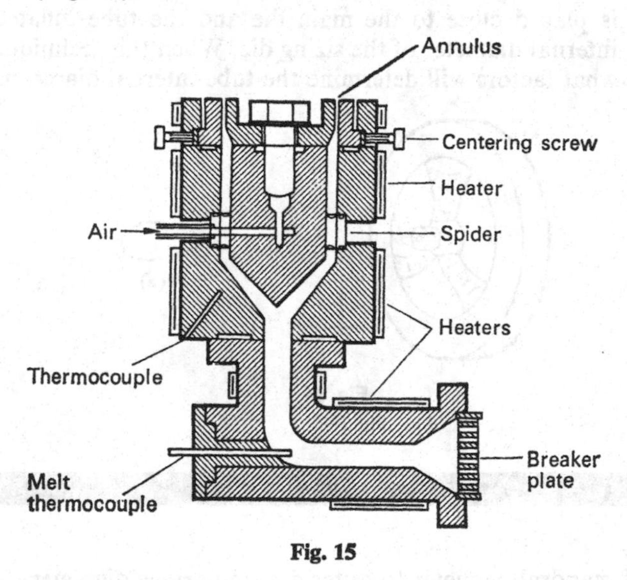

Fig. 15

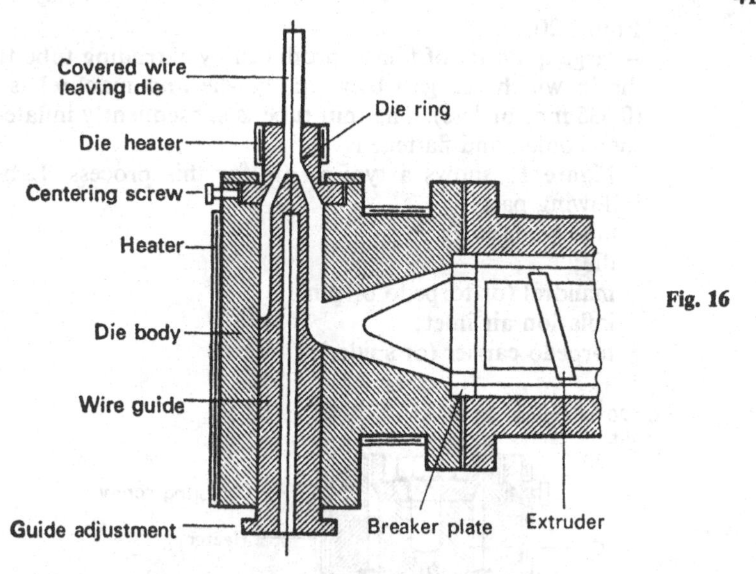

Fig. 16

Frame 18.
In Fig. 16 the torpedo (pin, mandrel) is hollow to enable wire to pass through. The die is thus suitable for wire covering. To ensure a concentric layer of polymer round the wire, centering screws act on the die which may be adjusted with respect to the ___.

pin (torpedo, mandrel)

Frame 19.
Note that the flow of the melt is broken by the presence of the torpedo, the split melt rejoining on the far side of the torpedo from the extruder barrel. To ensure a good fusion the die should be at a sufficiently high temperature and there should be restrictions on the die exit. This can be achieved by keeping the gap between the torpedo tip and die small and by ___ the length of the die parallel.

increasing

Frame 20.
A large quantity of film is produced by extruding tube from a die in which the gap between the die and mandrel is small (0·635 mm or less). The thin tube is subsequently inflated and then cooled and flattened.

Figure 17 shows a typical die for this process. Label the following parts:

elbow adaptor;
die;
mandrel (or torpedo or pin);
inflation air inlet;
torpedo carrier (or spider).

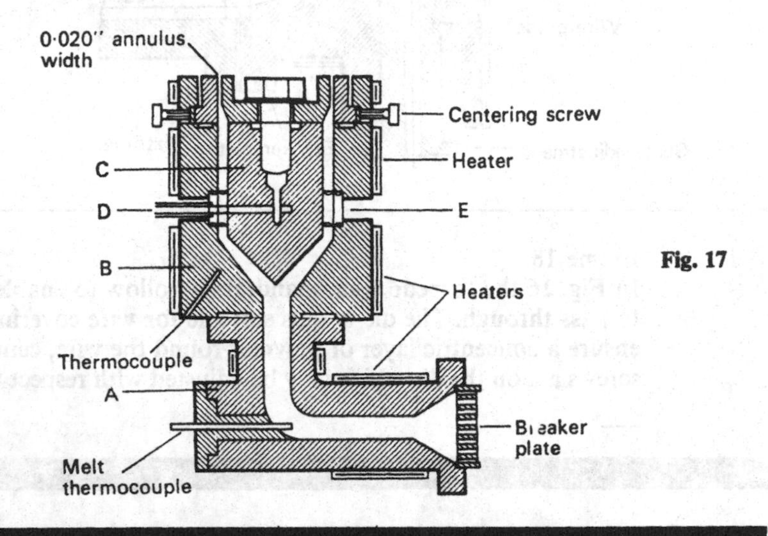

Fig. 17

A—elbow adaptor; B—die; C—mandrel; D—inflation air inlet; E—torpedo carrier

Frame 21.
How is an even film thickness obtained?
(a) by adjusting the mandrel with respect to the die?
or
(b) by adjusting the die with respect to the mandrel?

(b)

Frame 22.
The restrictor bulge is to help consolidation of melt which has been split on passing through the holes of the ________ ________, a sort of robust spider.

torpedo carrier

Frame 23.
It is important in film extrusion that the melt flows out of the die at the same velocity all round the gap. This means that it is necessary to ensure that melt is delivered evenly to all parts of the die. A spiral mandrel die (Fig. 18) is of use here.

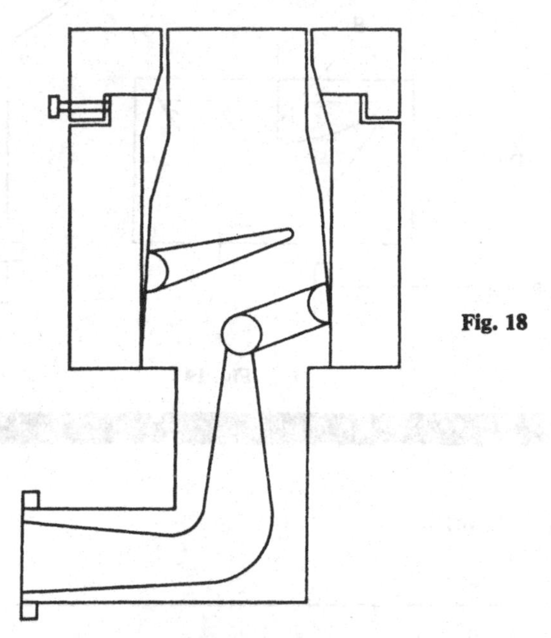

Fig. 18

The melt flows from the adaptor through a hole to a spiral channel around the mandrel. As the melt flows up the spiral, progressively more melt is able to escape from the spiral and flow by the annular gap between mandrel and die. Eventually all the melt is flowing up the annulus and on into a reservoir just behind the die parallel.

A spiral mandrel die helps to ensure an _____ distribution of melt all round the die gap.

even

Frame 24.

The extruder is often used to make plastics materials into sheet form (it is usually taken that the term sheet is used where the thickness exceeds 0·025 cm and the term film for thinner products). A die is normally used here in which the polymer emerges from a long flat slit. Whatever detailed design of die is used it is important that the molten polymer flows out of the die at a constant velocity and thickness all along the die lips.

Figure 19 shows a typical fish tail type of die.

Label the following parts:
(a) adjustable die lip;
(b) die lip adjusting screws;
(c) die parallel at die centre.

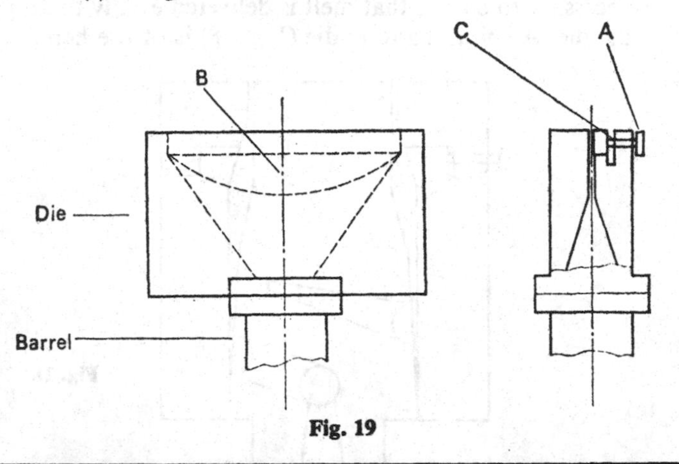

Fig. 19

(a) C; (b) A; (c) B

Frame 25.

In a fish tail die the parallel is longer in the die centre than at the sides to provide a restriction to flow. If this were not done (and assuming constant die gap) then the melt would flow faster/slower out from the die centre.

faster

Frame 26.

Fish tail dies are not desirable for making wide sheets for three of the following four reasons. Which one of these do you think is not a valid reason?

(a) they are very bulky;
(b) very high clamping forces are required to hold the die halves together;
(c) it is neither possible to vary the die gap nor to adjust the temperature along the die lips;
(d) the energy requirements to heat the large surface of such dies become economically prohibitive.

(c)

Frame 27.
Figure 20 shows a typical 'manifold die' design which is widely used for making sheet.

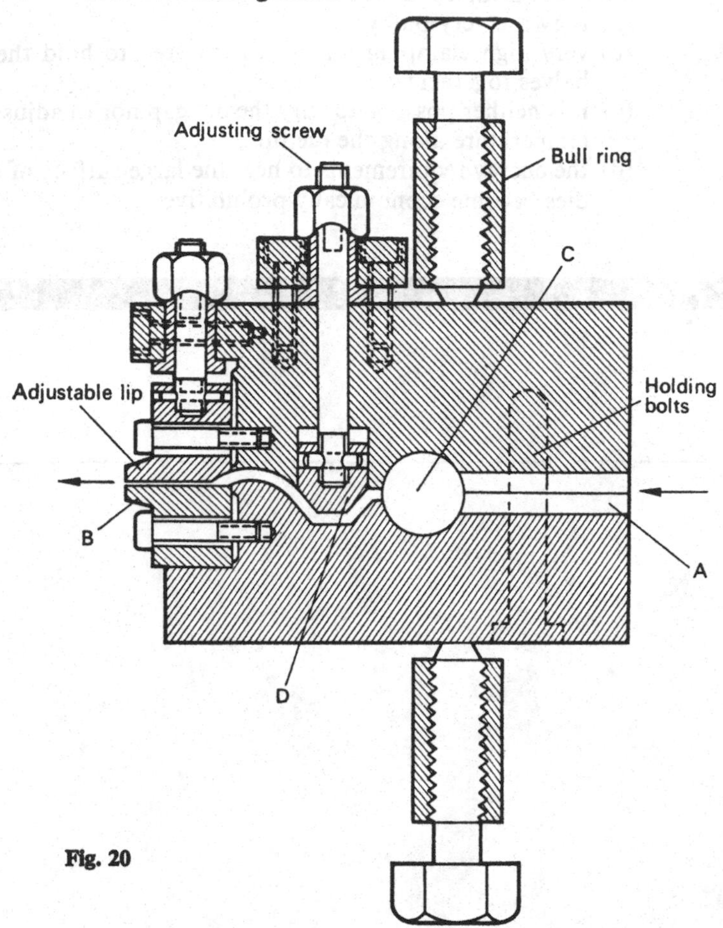

Fig. 20

The polymer is pumped from the extruder to the manifold (a long channel down the length of the die) *via* a short inlet tube. The melt flows along the manifold and then past a restrictor bar and adjustable die lips out of the die. Because of a drop of pressure occurring in the melt in the manifold, as the distance increases from the inlet tube the flow rates from the die tend to be greatest near to the inlet tube.

For this reason do you think that the inlet tube should feed to the die centre or to one end of the die?

---

to the die centre

Frame 28.

Feeding to the die centre reduces, but does not eliminate, the different flow rates. It is therefore common to fit an adjustable restrictor bar behind the die lips to throttle or otherwise adjust the flow of polymer to the die lips.

Label the following items on Fig. 20:

(a) fixed die lip;
(b) adjustable restrictor bar;
(c) inlet tube;
(d) manifold.

(a)B; (b) D; (c) A; (d) C

Frame 29.

For viscous melts, restrictor bars alone are often insufficient to compensate for the pressure drop along the die. One way of reducing the problem is to 'bend' the manifold so that the length of the die parallel is greater near the die inlet and tapering off towards the die ends. This arrangement is sometimes known as a coat-hanger die, and is shown in Fig. 21.

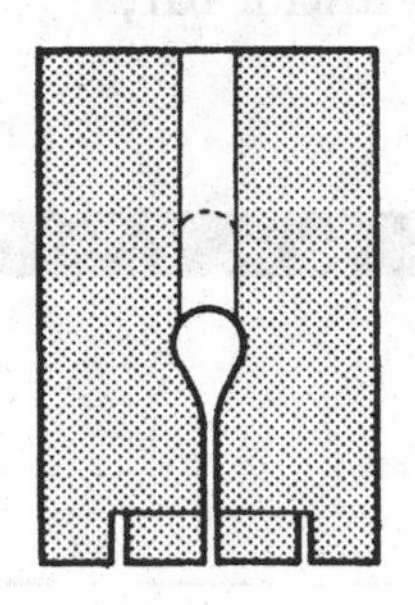

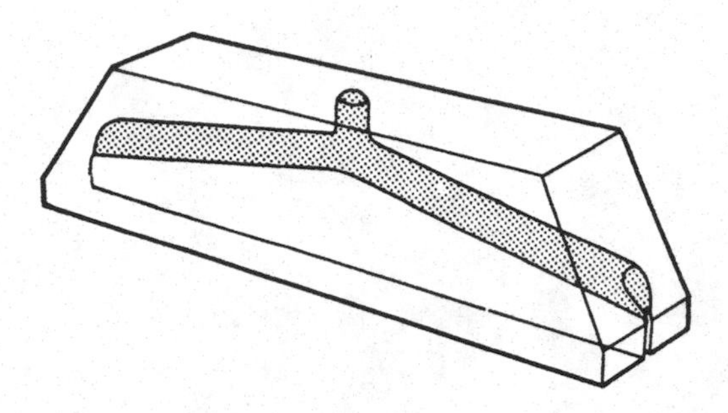

Coat-hanger' sheet extrusion die.

Fig. 21

A coat-hanger die is therefore of use to __________ for the pressure drops in a manifold with _______ melts.

---

compensate; viscous

Part Three

1. What happens to the long chain-like polymer molecules when they are sheared in a die?

2. How may die swell be reduced without redesigning a die or reducing extrusion rate?

3. How is pressure built up in the melt after it has passed round the spider legs and through the breaker plate?

4. Why is it so important to build up this pressure?

5. Name one type of die which may be used for making film by the tubular process and which helps to ensure that the melt flows out of the die at the same velocity all round the gap.

6. Why are manifold dies usually preferred to fish tail dies for the manufacture of sheet?

PART FOUR

MATERIAL CHARACTERISTICS AND COMMON EXTRUSION PROBLEMS

Frame 1.

Most thermoplastics materials are extruded on a commercial scale. The most important in terms of tonnage are:

polyvinyl chloride (p.v.c.);
polyethylene and polypropylene;
polystyrene and high impact polystyrene;
ABS (polystyrene complexes based on acrylonitrile, butadiene and styrene);
polymethyl methacrylate;
nylons (polyamides).

Frame 2.

Some thermoplastics absorb small amounts of water and they are said to be hygroscopic. In an extruder this water turns to steam and the resulting extrudates contain bubbles formed by the steam causing the extrudate to expand as it emerges from the die.

It is most important in these cases that the material should be quite _____ before extrusion is attempted.

dry

In the above list of polymers the problem can be most acute with the nylons and with polymethyl methacrylate.

Frame 3.

What modification to an extruder may be made so that it can handle slightly damp materials?

a screw with a devolatilising zone could be fitted (Part Two, Frame 21)

Frame 4.
Sometimes the bubbles caused by the steam become flattened during shear. These flattened bubbles tend to reflect light in transparent extrudates and appear like mica-flakes and are referred to as mica marks.
Can you suggest any other cause of such marks?

entrapped air bubbles; other volatile liquids including plasticisers; monomer formed on degradation of polymer

---

Frame 5.
The amount of heat required to bring a thermoplastic to its processing temperature will depend on:
(a) ....;
(b) ....

(a) the processing temperature
(b) the specific heat of fusion
If this question was answered correctly go on to Frame 6. If not attempt Frame 5*.

Frame 5*.
The specific heat of a material is the amount of heat required to raise unit quantity of the material one degree in temperature. The amount of heat is usually expressed in calories, the quantity in grams and the temperature in degrees Centigrade (Celsius).
A calorie is the amount of heat required to raise one gram of water one degree Centigrade. What then is the specific heat of water?

1·00 calorie per gram per degree Centigrade

Frame 6.
If the average specific heat of a grade of unplasticised p.v.c. is 0·24 calories per gram per degree Centigrade and of plasticised p.v.c. 0·40 and both are to be processed at 170°C, which will require the most heat input per gram to bring it to processing temperature?

plasticised p.v.c.

Frame 7.
Some thermoplastics are somewhat crystalline in nature. In order to melt the crystals latent heat will need to be supplied in addition to the heat required to raise the temperature.

In the case of polyethylene it has been estimated that a typical value for specific heat is 0·55 cal/g/°C and for latent heat 50 cal/g. How much heat per gram will need to be supplied to raise the melt from 20°C to a processing temperature of 220°C?

$0{\cdot}55 \times (220 - 20) + 50 = 160$ cal

Frame 8.
With polystyrene there are no crystals to melt and therefore no latent heat of fusion. If the specific heat is 0·32, how much heat has to be supplied to raise it by the same amount as in the previous example?

$0{\cdot}32 \times (200) = 64$ cal

Frame 9.
Polymer melts differ enormously in their viscosities. In addition, viscosity will vary with the rate of shear and the temperature of the melt.

In order to obtain a consistent final melt viscosity it is important that careful control of (a)__________ and (b)_____ _____be maintained.

(a) temperature; (b) screw speed (controlling rate of shear)

Frame 10.
The higher the melt viscosity the_______the power consumption.

higher (greater)

Frame 11.
Some polymers are very fluid in the molten state. Extrudates from such materials tend to distort unless they can be solidified very quickly.

Frame 12.
It is found that the melt viscosity increases rapidly with increase in the molecular weight of the polymer.

Hence when there are problems handling fluid polymers it is often advisable to change to a higher/lower molecular weight grade of polymer.

higher

Frame 13.
Polymers degrade on subjection to continual heating, but some are more resistant than others. Among the important commercial thermoplastics, p.v.c. is probably the most susceptible to degradation since it is processed at temperatures close to its decomposition temperature. Therefore when extruding p.v.c. very accurate control of ____________ will be necessary.

temperatures (of the melt, of the barrel)

Frame 14.
It is important that there are no places in the extruder or the die head (dead spots) where molten polymers can stagnate and decompose.

Frame 15.
One of the advantages of thermoplastics is that faulty mouldings, extrudates and the scrap material can be ground up and reprocessed. This can only be carried out a very limited number of times with p.v.c. because of the risk of ____________.

decomposition

Frame 16.
On emerging from the die of an extruder a thermoplastics material is in the molten state. Whilst cooling and hence setting it must be held in shape to prevent it __________.

distorting

Frame 17.

Figure 22 shows a typical arrangement for making a simple section. The extrudate is carried away from the die by means of a conveyor belt. If the extrusion rate is kept constant and the conveyor belt speed increased, what will happen to the extrudate cross section?

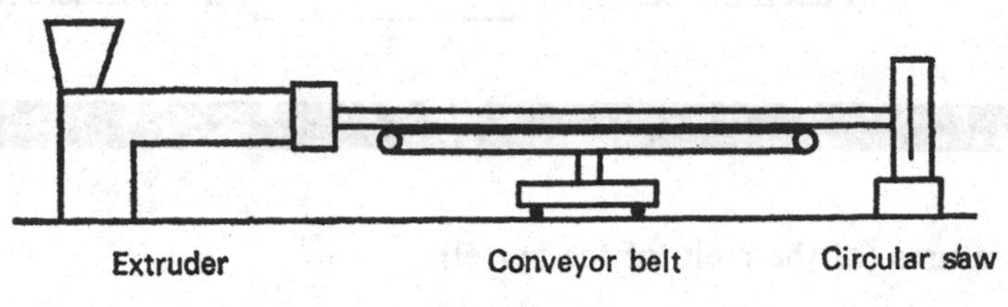

Fig. 22

---

it will decrease

Frame 18.

Air cooling is slow. Can you suggest a means of increasing the rate of cooling?

---

cooling in water or spraying with water

Frame 19.
Figure 23 shows an arrangement with a water cooling trough. In this case the haul-off rate is controlled by a foam rubber-clad double-caterpillar band device. Sizing plates are fitted in the water bath to ensure that the extrudate is being drawn down to the correct extent.

Use of water cooling will_______ the cooling time compared with the time required when air cooling the same extrudate.

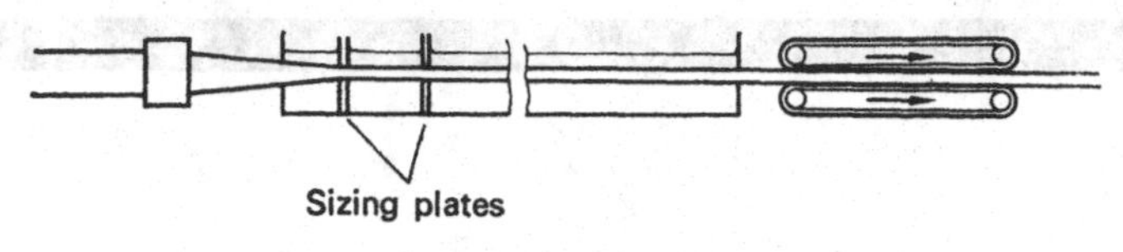

Fig. 23

---

reduce

Frame 20.
Section extrudates have been made from most thermoplastics. In general, if the extrudate has a smooth but matt surface when a glossy finish is required, this indicates that the melt generally has been suitably heated but that the die temperature is too cold. Therefore increased gloss (at least with many thermoplastics) may often be obtained by__________ the die temperature.

---

increasing (but *see also* Frame 13)

Frame 21.
Some extrudates may have a glossy finish but are lumpy and very irregular. This is usually the result of poor mixing of the melt. Good mixing is necessary because in the extruder barrel layers of melt will be at different temperatures and also subjected to different shear conditions and will thus have different __________.

viscosities

Frame 22.
The degree of mixing of the melt may be increased by increasing the back pressure on the melt in the barrel.

Give two ways of increasing the back pressure by changes in the die head design.

increasing the die parallel; fitting stronger screen packs (*i.e.* the mesh should be finer or there should be several layers of mesh)

Frame 23.
Decreasing the melt temperature in the die will increase melt viscosity and increase the back pressure. Hence decreasing the die head temperature may help to ______ lumpiness.

reduce

Frame 24.
A die is being used to produce a rectangular section extrudate which shows slight lumpiness. It has been suggested that increasing the cross section of the die (and subsequently increasing the drawing down to bring the product to size) will help to reduce lumpiness. Is this true or false?

false—increasing the cross section will reduce the back pressure and is liable to aggravate the situation

---

Frame 25.
Extrudates sometimes show bubbles and porosity. This is due to trapped gases. These can arise through air trapped between granules which have become surrounded by molten polymer, by vaporising of additives such as plasticisers, by absorbed water, by depolymerisation yielding volatile monomer or polymer degradation giving gaseous by-products (*e.g.* hydrochloric acid from p.v.c.).

Name one method described in Part Two by which volatiles may be removed before the melt reaches the die.

the use of a devolatilising screw (screw with a decompression zone)

Frame 26.
Blistering may be caused by water either absorbed into the granules or lying on the surface. The amount of steam produced to give blisters will depend on (a) the amount of water present and (b) the extrusion temperature. Hence polymers which are hygroscopic and/or require a high process temperature are normally carefully dried and stored in sealed tins.

The amount of blistering will_________with the amount of water present and will increase with the extrusion temperature.

increase

Frame 27.
The likelihood of depolymerisation causing blistering, etc. can be reduced by using a ________ temperature profile along the barrel.

flatter

Frame 28.
Extrudates sometimes exhibit a surface irregularity characterised by a series of ridges perpendicular to the flow direction. This is known as sharkskin and is often barely visible to the naked eye but results in a form of mattness or lack of surface gloss.

For a particular material extruding at a given temperature the sharkskin appears to occur above a critical linear extrusion rate independent of the die cross-sectional area. It is believed to be due to a stick–slip effect in the die exit region. The critical extrusion rate appears to increase with increase in temperature. Hence sharkskin tends to be ________ with increase in temperature.

reduced

Frame 29.

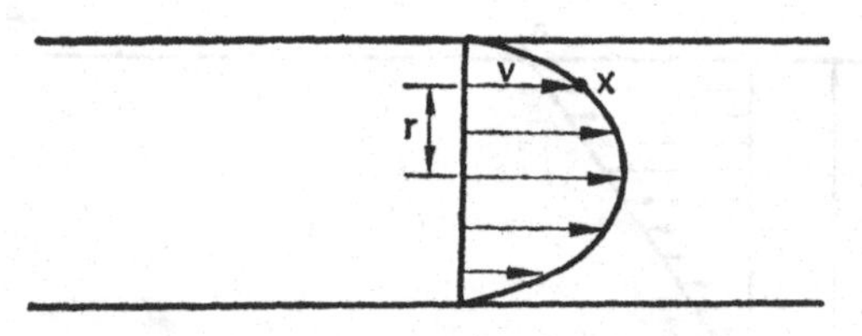

Fig. 24

In Fig. 24 we have a velocity profile for flow in a channel. At a distance r from the centre the velocity of the liquid is v. What term is given to the slope of the profile at point X (*i.e.* dv/dr in calculus terminology)?

---

the shear rate
If you obtained the correct answer go to Frame 30, if not go to Frame 29*.

Frame 29*.

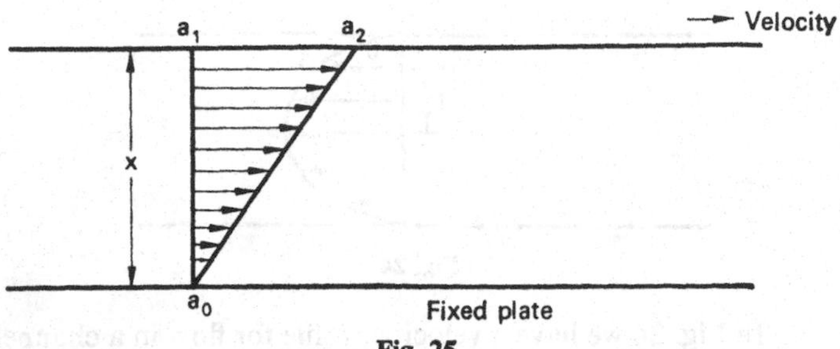

Fig. 25

Consider two parallel plates distance x apart, the space between being filled by a liquid. The lower plate is fixed and the upper one moves at a velocity v in the direction shown, under a force F which is applied to the area A of the plate (it is assumed that A is very large compared with x). This will cause particles of liquid which are at one time along the line a_1–a_0 to take up a position along the line a_2–a_0 at some later time, assuming that there is no movement of particles at the wall. The arrows in the y-direction indicate the relative velocities of the particle.

The way in which velocity v changes with distance x from the bottom plate (dv/dx in calculus terminology) is known as the shear rate. In this instance the shear rate is equal to v/x. The force per unit area (F/A) is known as the shear stress.

How does the shear rate change with position between the two plates?

---

it does not change, it is constant

Frame 30.

What name is given to a liquid in which the ratio shear stress/shear rate is a constant known as the coefficient of viscosity?

Newtonian fluid

If you obtained the correct answer go to Frame 31, if not go to Frame 30*.

Frame 30*.
For many simple fluids like water the ratio shear stress/shear rate is constant. Such materials are said to be Newtonian. In other cases the ratio changes with increased shear (*see* Fig. 26).

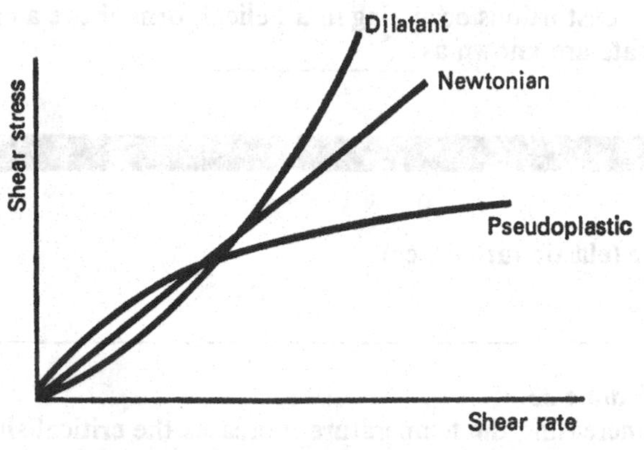

Fig. 26

In general the ratio shear stress/shear rate is known as the apparent viscosity. Therefore pseudoplastic materials become more/less viscous with increase in shear rate.

less

Frame 31.
Thermoplastic melts almost invariably show pseudoplastic behaviour. Thus the apparent viscosity of polyethylene ________ with increased shear rate.

decreases

Frame 32.
Above some critical shear rate most thermoplastics exhibit a phenomenon variously known as elastic turbulence or melt fracture. It is characterised by various types of distortion which have a helical form.

Distortions occurring in a helical form above a critical shear rate are known as _____ ________.

---

melt fracture (elastic turbulence)

Frame 33.
Increasing the temperature increases the critical shear rate for onset of melt fracture.

The effect may therefore be reduced and perhaps even eliminated by reducing the shear rate and/or __________ the temperature.

---

increasing

Frame 34.
The critical shear rate for onset of melt fracture is reduced by increasing molecular weight. Therefore melt fracture is more/less likely to occur with high molecular weight grades than low molecular weight grades.

---

more

Frame 35.
Give those ways by which melt fracture effects may be reduced.

(a) lower the shear rate
(b) raise the temperature
(c) use a material of lower molecular weight

Frame 36.
The extruder may show black lumps or flecks. These may be due to polymer stagnating at some points in the machine and decomposing. Pieces of decomposed material are then swept away by molten polymer at irregular intervals.

The problem may be avoided by:

(a) lowering extrusion temperature;
(b) regularly cleaning dies;
(c) avoiding dead spots.

How else might black particles occur in light coloured extrudate?

it might be contaminated with some black compound

Frame 37.
Figure 27 shows the pattern of gelation in a barrel.

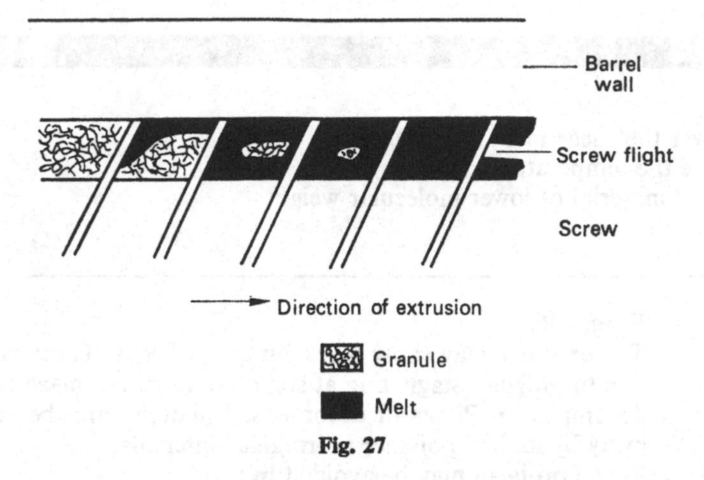

Fig. 27

It will be seen that by the time the last of the polymer is melting some granules are contained in an envelope of melt already formed. Air can only escape down a helical path through the granules back towards the hopper. If the rate of extrusion is greater, then the rate of back flow of air will be carried through into the extrudate resulting in porosity.

Porosity may therefore be reduced by ________ the rate of extrusion.

reducing

Frame 38.
The back flow of air may be increased by increasing the die head pressure. Suggest one way of doing this without making any changes to the die head or extruder design.

lower the head temperatures (to increase the melt viscosity)

Frame 39.
Porosity is usually most serious when feeding compound in powder form as it is often done with rigid p.v.c. compounds. Hence the critical extrusion speed for onset of porosity is often ______ with powder feed than with granulated compound.

lower

Frame 40.
When polymer melts pass round a spider leg and/or through a breaker plate the molten material is separated. Unless there is a high pressure on the side of the spider and/or breaker plate the separated melt will not fully knit together and lines or planes of weakness will occur.

The head pressure may be increased by:

(a) increasing the die parallel;
(b) reducing the cross-sectional area at the die;
(c) lowering the die temperature.

Which of these adjustments may be made without removing the die head from the machine?

(c) lowering the die temperature

Frame 41.
When extruding polymers, additives are sometimes deposited from the melt onto the forward part of the screw and onto the extruder head and die. This effect is known as plate-out and is most frequently met with in p.v.c. Deposition occurs most frequently in regions of high temperature and high shear. Formulation also has an important influence.

Give three possible ways in which plate-out may be reduced.

(a) lower the die-head temperature
(b) reduce extrusion speed
(c) change the formulation

CRITERION TEST

Part Four

1. What are the most likely causes of mica marks in extrudates in the following instances?

   (a) carefully dried unplasticised p.v.c. extruded under conditions which did not cause significant polymer degradation;
   (b) carefully dried polymethyl methacrylate extruded from an extruder with a high compression ratio screw.

2. Which polymer usually requires more heat to bring it to its processing temperature, plasticised p.v.c. or polyethylene?

3. What are dead spots and why should they be avoided?

4. Thermoplastics are usually dilatant—true or false?

5. Give one approach which may eliminate melt fracture in an extrudate without changing the grade of thermoplastic used or without reducing output rate.

6. Is porosity more likely with a powder feed or a granule feed?

PART FIVE

THE COMPLETE EXTRUSION PROCESS

As was indicated in Part Four when discussing the extrusion of section, the extruder and die form only a part of the whole process of making products by extrusion. In this part we shall consider selected important products and show how the extruder is used in order to make them.

EXTRUDED PIPE

Frame 1.
A typical die head for making extruded pipe was shown in Part Three (Fig. 13).

Figure 28 below shows a typical layout for the complete extrusion plant for making rigid piping from, for example, unplasticised p.v.c.

Label the following parts on the diagram:

(a) sizing die;
(b) haul off;
(c) water tank;
(d) circular saw;
(e) floating plug.

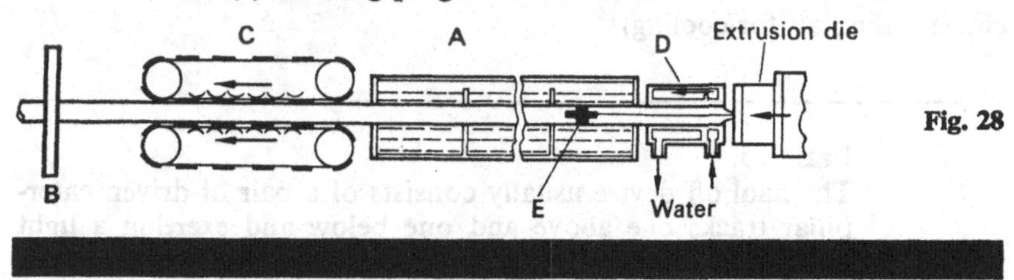

Fig. 28

(a) D; (b) C; (c) A; (d) B; (e) E

Frame 2.
The most common way of determining the final external dimensions of the pipe is by means of a sizing die. This is a cylindrical, water-cooled jacket whose internal bore is identical to that of the desired external pipe dimensions. It is located close to the die face and is of slightly greater diameter than the die orifice. By means of compressed air introduced through the torpedo, the pipe may be inflated to the size of the die. What do you think is the function of the floating plug?

to prevent air escaping out of the end of the pipe

Frame 3.
Can you think of any other way of bringing the pipe up to the size of the sizing die?

by applying a vacuum through holes in the wall of the sizing die; this method is normally only suitable for small bore tubes of diameter less than 2 cm but can be used on large diameters if special means of sealing between the die and the tube, *e.g.* flexible sealing rings, are used.

Frame 4.
What is the function of the water tank?

to cool the extrudate before it reaches the haul off device (water is more efficient than air for cooling)

Frame 5.
The haul off device usually consists of a pair of driven caterpillar tracks one above and one below and exerting a light pressure on the pipe.

With large pipes two tracks often tend to cause distortion and do not always give a uniform pull. Can you suggest how this might be overcome?

by fixing another pair of tracks, one on each side of the pipe

---

Frame 6.
With small diameter tube, for example for ball point pen refills, it would be very difficult to use a sizing die. Can you suggest a way of ensuring that the extrudates are of the correct size?

by extruding slightly oversize and pulling off at a greater rate than the melt is extruding; the diameter can be checked by the use of templates

---

FILM

Frame 7.

Figure 29 shows a typical arrangement for making film by the tubular process. The polymer is extruded vertically through a die (such as that shown previously in Fig. 17) into a thin tube which is then inflated with air, cooled, flattened between rollers and wound up. Label the following parts on Fig. 29:

(a) die;
(b) cooling ring;
(c) nip rolls;
(d) wind up.

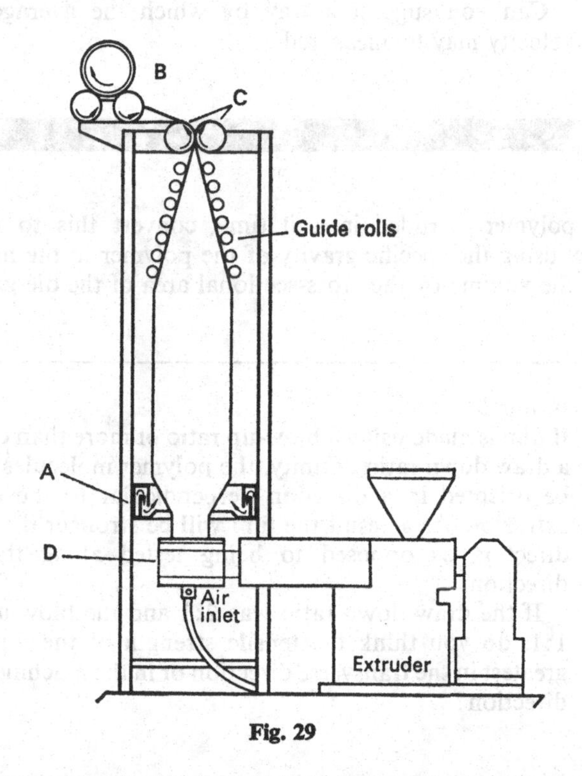

Fig. 29

(a) D; (b) A; (c) C; (d) B

Frame 8.

Although suitable for many thermoplastics, the tubular process is used mainly for making large quantities of polyethylene film.

Two operational terms need to be defined:

(a) the blow-up ratio—this is the ratio of the diameter of the bubble to the diameter of the die annulus. In practice blow-up ratios between 2:1 and 6:1 are most commonly used;

(b) the draw down ratio—this is the ratio of the velocity of the film through the nip rolls compared to the velocity of extrusion.

Can you suggest a way by which the average extrusion velocity may be measured?

---

weigh the polymer extruded in unit time, convert this to the volume extruded by using the specific gravity of the polymer in the molten state, and divide the volume by the cross-sectional area of the die gap

---

Frame 9.

If film is made using a blow-up ratio of more than one but with a draw down ratio of unity, the polymer molecules will tend to be oriented in a direction perpendicular to the direction of extrusion. As a result the film will be stronger if tested in this direction as opposed to being tested along the extrusion direction.

If the draw down ratio was 2:1 and the blow-up ratio was 1:1, do you think the tensile strength of the film would be greatest in the transverse direction or in the machine (extrusion) direction?

---

in the extrusion direction

Frame 10.
When extruding polyethylene by the tubular process it is observed that immediately above the die the melt is quite transparent. However, about 20 cm above the die the polyethylene is seen to develop haze. The distance above the die that this haze develops is known as the freeze line distance (or frost line distance). This is due to a form of crystallisation developing on cooling of the melt.

Frame 11.
If all other operating conditions were kept constant, how do you think the frost line distance would change with an increase in melt temperature?

it would increase as the polymer would take longer to cool down to its crystallising temperature and so the film would have travelled further from the die

---

Frame 12.
If all other operating conditions were kept constant, how do you think the frost line distance would change with an increase in extrusion rate (and haul off rate—so as to keep the draw down ratio constant)?

it would increase

Frame 13.
It is commonly found that for a given grade of polyethylene and at specified operating conditions the haze of film goes through a minimum with change in frost line distance. It is therefore important to carefully control____ ____ ________ in order to minimise haze.

frost line distance (it is in fact important to control all operational variables)

Frame 14.

When film is stretched in the melt the molecules tend to orient. If, however, the film is kept molten after it has been stretched, the molecules will tend to slide past each other and dis-orient, the amount of dis-orientation increasing with time.

Since the lateral stretching (related to the blowing up of the tube) does not occur in quite the same way and at the same time as the longitudinal stretching due to drawing down, tube made using conditions of a 2:1 blow-up ratio and 2:1 draw down ratio does/does not have equal orientation in the longitudinal and transverse directions.

does not

Frame 15.

The relative orientation may be affected by the shape of the bubble below the frost line and this can be controlled by such operating conditions as extrusion speed and cooling rate.

It is therefore often possible to obtain a film with balanced orientation by altering the ______ of the bubble.

shape

Frame 16.
In Fig. 30(a) the polymer molecules are shown schematically aligned in the longitudinal direction and in Fig. 30(b) aligned in the transverse direction.

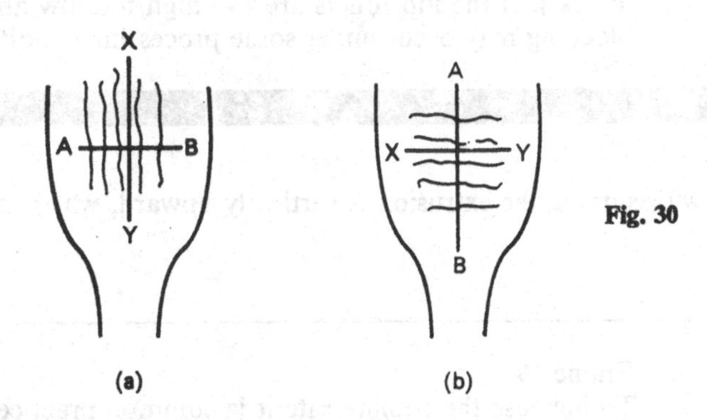

Fig. 30

The film in each case is much stronger in the XY direction than in the AB direction.

If some evenly tensioned film is subjected to an impact blow perpendicular to the surface of the film sufficient to rupture the film, do you think that it would tear along a line (a) parallel to AB or (b) parallel to XY?

(b) parallel to XY

Frame 17.
Do you think that the highest impact strength will be obtained:
- **(a) when the orientation is greatest in the longitudinal direction;**
- **(b) when the orientation is greatest in the transverse direction;**

or
- **(c) when the orientation is balanced?**

(c) when the orientation is balanced (in the other two cases there are weak planes parallel to the molecular orientation)

Frame 18.
It is important that the film is sufficiently cooled before it is flattened out by the nip rollers. If this is not done the two surfaces of the film will be difficult to separate (will tend to 'block'). If the nip rollers are too high/too low above the die, blocking may occur under some processing conditions.

too low (assuming the extrusion is vertically upward, which is usually the case)

Frame 19.
To increase the cooling rate it is common practice to impinge a stream of cool air evenly around the bubble from a cooling ring.

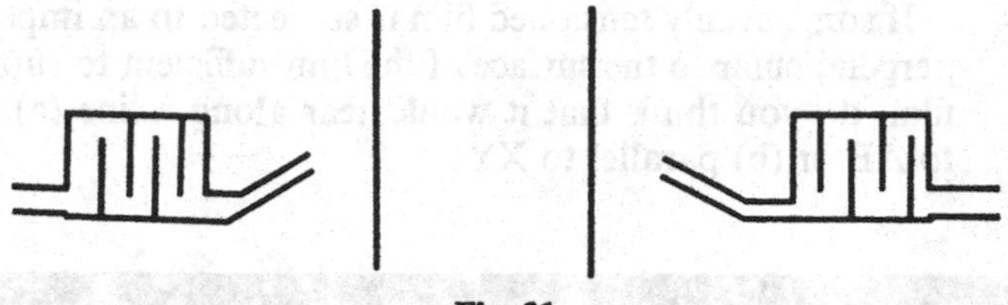

Fig. 31

A cooling ring is therefore used to ______ the time taken for the polymer to freeze.

reduce (shorten)

Frame 20.
Air is an inefficient heat transfer medium. For many films extruded with a fixed position of the nip rolls above the die and extruding film of specified dimensions the factor limiting the maximum rate of output is the rate of cooling in ___.

air

Frame 21.
In order to achieve faster output rates the cast-film method of extrusion is sometimes employed. This is shown schematically in Fig. 32.

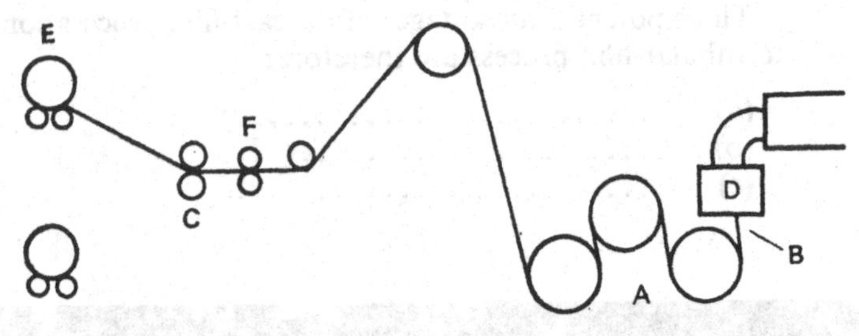

Fig. 32

However, whereas film several metres in diameter may be made by the tubular process, it becomes difficult to work the cast-film process for sheet greater than about 3 m wide. If sheet is being made 15 m wide by the tubular process and 3 m wide by the cast-film process, how many times faster must be the haul off rate in the cast-film process compared to the tubular process in order to achieve the same area output per unit time?

five times

Frame 22.
Label the following parts in Fig. 32:
(a) die;
(b) air gap;
(c) chill rolls;
(d) rubber covered nip rolls;
(e) trim rolls;
(f) wind up roll.

(a) D; (b) B; (c) A; (d) C; (e) F; (f) E

Frame 23.
In making polyethylene film by the cast-film process it is possible to use high temperatures and high cooling rates. This can reduce surface irregularities in the film and hence reduce haze and increase gloss.

Three potential advantages of the cast-film process compared to tubular-film process are therefore:

(a) ....;
(b) ....;
(c) ....

(a) lower haze of film
(b) greater gloss of film
(c) increased extrusion rate

Frame 24.
The chill rolls are controlled in temperature by circulating water. For low density polyethylene the chill rolls are usually kept at about 50–70°C. Since the die gap is usually about 0·5 mm to produce film about 0·025 mm, the draw down ratio is about ____.

20:1 (ignoring lateral contraction of the film)

Frame 25.
The draw down ratio is controlled by adjusting the extrusion speed and the speed of the ____ rolls.

---

chill

Frame 26.
Considerable molecular orientation occurs during the drawing down of the melt. This can result in low film strengths, particularly in the direction parallel to the extruder direction.

Molecular orientation in the machine direction will decrease with the air gap distance and so an increase in the air gap distance may________ the strength properties.

improve

Frame 27.
Very short air gaps may also increase the tendency of the melt to tear due to the sudden acceleration effect and also for surface irregularities in the melt not to have time to even out thus producing optical irregularities.

Three reasons, therefore, for increasing the air gap distance are:

(a) ....;
(b) ....;
(c) ....

(a) to reduce molecular orientation
(b) to reduce the tendency of melt tearing
(c) to reduce optical irregularities

Frame 28.
If, however, the air gap is too great, some air cooling occurs which allows haze to develop through the slower crystallisation. In addition the molten web tends to contract laterally reducing the width of film and increasing the amount of edge bead which has to be trimmed.

For a given polymer being extruded at a given temperature through a given die it is necessary to compromise on the ____ ____ distance.

---

air gap
In a typical low density polyethylene process extruded with a melt temperature of 300°C, an air gap distance of some 25–50 mm is commonly employed.

SOLID SECTION

Frame 29.
The extrusion of solid section is widely carried out to produce such objects as guttering, curtain runners, conduit and solid rod (for subsequent machining operations). In such cases the overall process is similar to that used for pipe except the problem no longer exists in blowing the pipe up to the correct size. Cooling baths and haul off arrangements are, however, of the same general type.

In some processes where the thermoplastics material goes through a broad temperature range in which it is rubbery it is possible to carry out useful post-forming operations during the extrusion process.

Post-forming is often possible when a thermoplastics material goes through a ________ state.

---

rubbery (elastic)

Frame 30.
Amongst polymers which are sometimes post-formed during the extrusion process are rigid p.v.c., polymethyl methacrylate and polystyrene. This is because they go through a ______ state on cooling.

rubbery

Frame 31.
Polymethyl methacrylate light fittings are often made in this way. Figure 33 shows a typical arrangement. The melt emerges from a serrated die being cut by a fixed knife at one point on emerging from the die. The extrudate is then drawn round a series of templates slowly being distorted to the desired shape. Air cooling is generally performed before the melt is drawn off by the haul off device and cut up into appropriate lengths.

Label the following parts on the figure:
(a) cutting knife;
(b) templates;
(c) air jets;
(d) haul off.

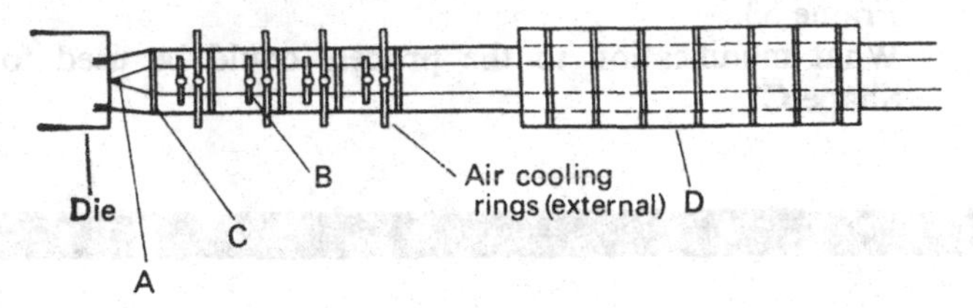

Fig. 33

(a) A; (b) C; (c) B; (d) D
(Note: the serrated die is only used to give a reeded appearance)

Frame 32.
Which of the following cross sections could be obtained by the technique described above?

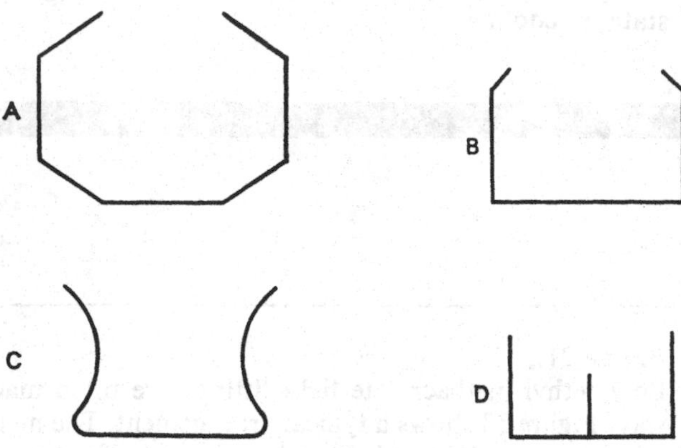

Fig. 34

A; B

Frame 33.
What modification to the process could be used to obtain Shape C?

the rubbery extrudate could be led between shaping rollers, *e.g.*

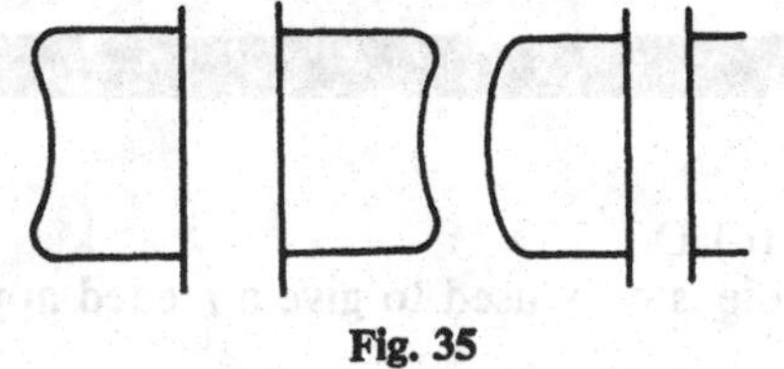
Fig. 35

Frame 34.
How may a diamond or other similar pattern be introduced on to the extrudate from a smooth (*i.e.* non-serrated) die?

by putting the rubbery extrudate between a smooth and a patterned roller under pressure

Frame 35.
How could an extrudate be made with the following cross section with white side walls and a black base?

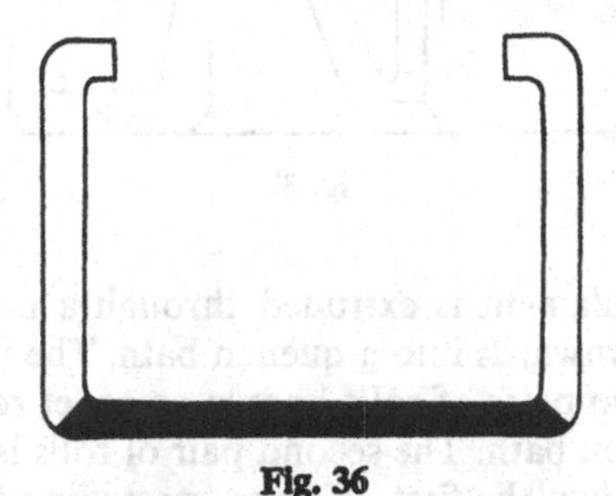

Fig. 36

by feeding two extruders into one die

MONOFILAMENT

Frame 36.
Extruded monofilament finds a variety of uses such as for the manufacture of filter cloths, garden furniture, radio grilles, fishing lines, racquet strings and brushes. The main thermoplastics used are polyethylene, nylon, polyvinylidene chloride and one amorphous polymer, polystyrene.

The other three polymers mentioned are __________ polymers.

crystalline

Frame 37.

The monofilament process consists essentially of three stages:

(a) extrusion into a quench bath;
(b) orientation;
(c) setting and cooling.

Figure 37 shows a typical arrangement for making high density polyethylene monofilament by a single-stage process.

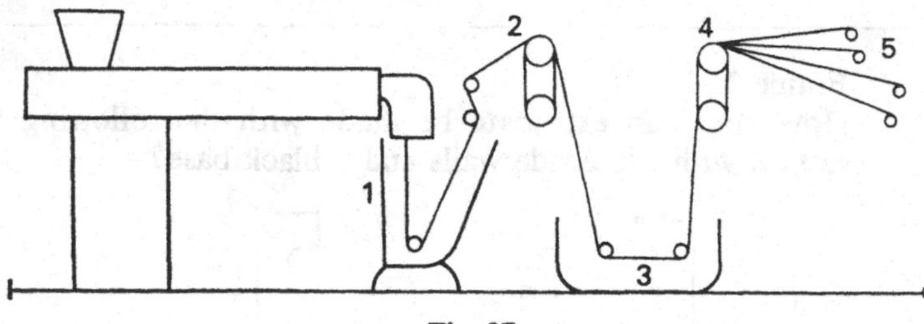

Fig. 37

Monofilament is extruded through a multi-holed die vertically downwards into a quench bath. The filaments are wound round two pairs of rolls known as godet rolls separated by an orientation bath. The second pair of rolls is rotated at a higher velocity than the first pair and, providing the monofilament is at a temperature above the glass transition temperature (a sort of softening temperature), the molecules and crystals are able to orient leading to a considerable increase in strength in the axial direction.

Label the following parts:

(a) first pair of godet rolls;
(b) orientation bath;
(c) final take off bobbins;
(d) quench bath;
(e) second pair of godet rolls.

(a) 2; (b) 3; (c) 5; (d) 1; (e) 4

Frame 38.
Nylon can be oriented at room temperature so that in this case there is no need for an__________ ___.

orientation bath

Frame 39.
In general both extrusion and orientation bath conditions have a critical effect on both the ease of extrusion and the final filament properties.

Amongst important factors to control are (with typical figures for high density polyethylene in parentheses):

(a) quench bath temperature (45–50°C);
(b) distance of quench bath from the die (<2·5 cm);
(c) degree of draw down between die and first pair of rolls (<5%);
(d) orientation bath temperatures (95–100°C);
(e) extrusion melt temperature (250°C).

For high density polyethylene, quench bath temperatures are thus______than orientation bath temperatures.

lower

Frame 40.
To ensure that all monofilaments emerging from the die have the same characteristics it is important that the polymer in each monofilament should have undergone the same amount of heating and shear in the extruder head and die. The die should therefore be designed so that all the holes are similarly disposed to the screw.

Which of the following dies do you think would be the most suitable?

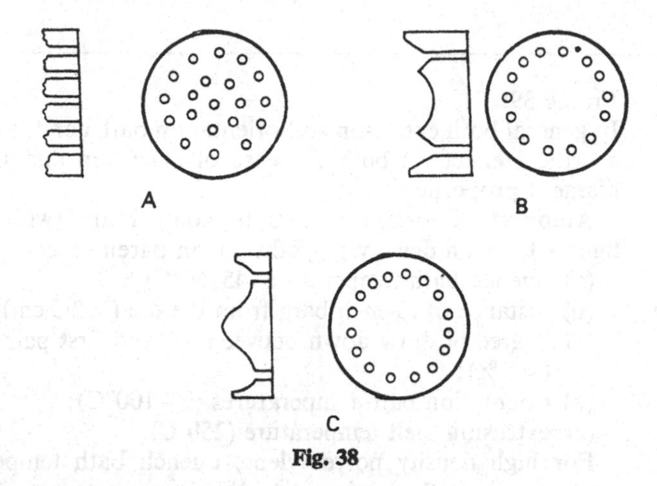

Fig. 38

C

Frame 41.
Can you suggest any other problems that might occur with die B?

---

dead spots would facilitate degradation (this can be a serious problem with monofilaments)

WIRE COVERING

Frame 42.
A wire covering die is similar to a die for making tube in that there is a torpedo present. The wire passes down the torpedo and is covered with molten polymer before it leaves the die head. A typical die head design is shown in Fig. 39.

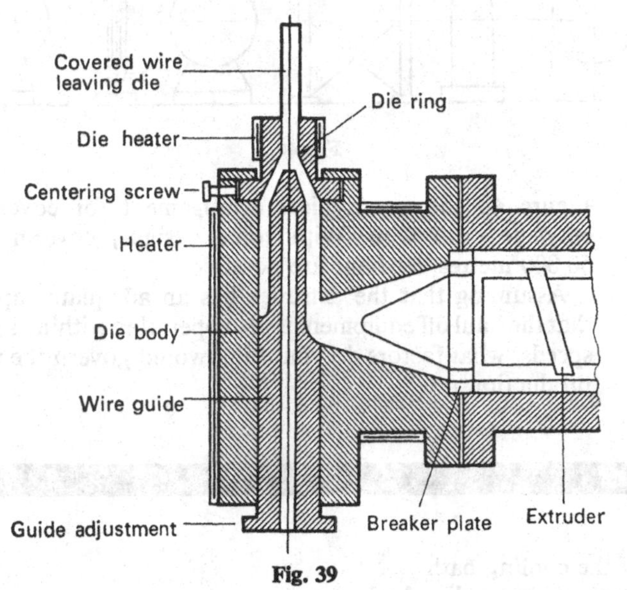

Fig. 39

Why do you think that the polymer is fed in at an angle to the direction of extrusion?

so that the wire does not have to pass through the extruder, which would introduce many complications

---

Frame 43.
In order to obtain a tight covering of polymer onto the wire the tip of the torpedo is positioned about 6 mm behind the die parallel.

If the tip of the torpedo extended beyond the die parallel there would be an undesirable ______ fit of the covering.

loose

Frame 44.

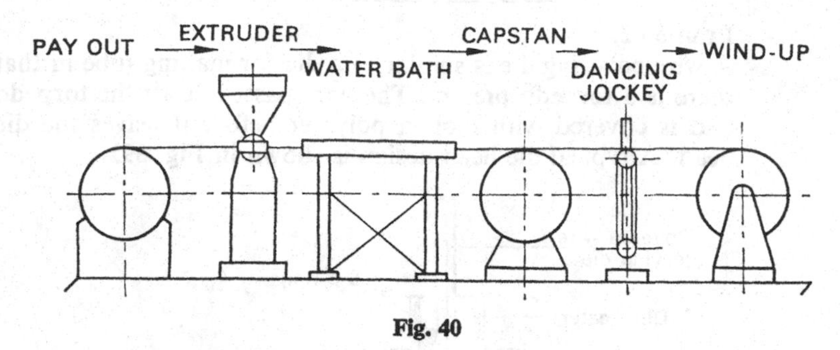

Fig. 40

Figure 40 shows a typical arrangement for covering polyethylene onto wire. With some designs, covering rates of 50 000 metres per hour are possible.

Assuming that the extruder has an adequate capacity and that the haul off equipment is also operating within its maximum speeds, what factors do you think would govern the maximum production rate?

---

length of the cooling bath
temperature of the cooling bath
type of polymer
thickness of the covering

Frame 45.
In order that the outer diameter of the covering should equal the die diameter the melt must be drawn down to compensate for the ______ on emerging from the die.

---

swell (die swell)

Frame 46.
Which of the following controls the actual linear covering rate?
(a) the screw speed;
(b) the capstan wheel;
(c) the wind-up drum.

(b) the capstan wheel

Frame 47.
It is important that the covering is concentric with the wire. For this reason, when setting up the equipment it is important to check that the die and ________ are concentric.

torpedo

Frame 48.
When covering with thick layers of crystallising thermoplastics such as polyethylene too fast a cooling rate may cause premature freezing of the outside of the covering. Because the external dimensions of the covering then become fixed further cooling shrinkage will be outwards, away from the conductor, and voids will be formed. This may adversely affect electrical insulating properties.

To prevent this from happening it is necessary to employ a more ________ cooling by passing the covered wire through a series of zones in the water bath which are at different temperatures.

gradual

Frame 49.
Under some extrusion conditions melt fracture might be observed during wire covering. In these circumstances the effect can be removed by__________the extrusion speed.

reducing (or increasing the temperature)

PART SIX

MULTI-SCREW EXTRUDERS

Frame 1.

The pumping action of the single-screw extruder requires that the feed material moves up the barrel without too much rotation (*see* Part Two, Frame 3). With some feed materials such as powders, pastes and low density polyethylene flake this requirement is not met. The pumping of these materials up the barrel is inefficient and it is said that the pumping is not positive.

Give an example of a form of material which does not give positive pumping in the few zones of a single screw extruder.

---

pastes; powders; low density polyethylene flake

Frame 2.

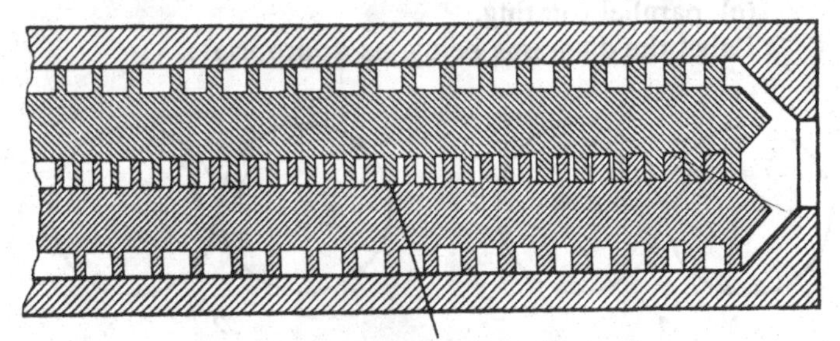

Fig. 41

In the 1930's multi-screw extruders were introduced in an attempt to provide a more positive acceptance of the feed material. Of these the most common are those with two screws only (twin-screw extruders).

The twin-screw extruders themselves can be classified into two types:

(a) extruders with screws intermeshed;
(b) extruders with screws not intermeshed.

In the figure shown the screws are __________.

---

intermeshed

Frame 3.

Intermeshing screws are more common because they give a more__________pumping action since the polymer is trapped into small C-shaped 'packets' and pushed along the barrel by the rotation of the screw.

positive

Frame 4.

Twin-screws may be:

(a) contra-rotating, in which the screw surfaces move towards each other below the feed throat;
(b) contra-rotating, in which the screw surfaces move away from each other below the feed throat;
(c) parallel rotating.

Label the diagrams below as appropriate.

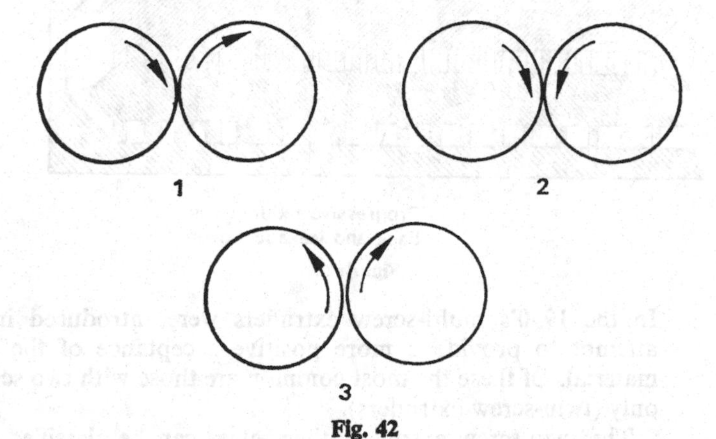

Fig. 42

(a) 2; (b) 3; (c) 1

Frame 5.
Which of the above designs would you consider most liable to damage if a hard foreign body was dropped accidentally down the feed throat?

---

2 (a)

Frame 6.
Non-intermeshing screws are not often used since they give a less ________ pumping action.

---

positive

Frame 7.
The diagram below shows the relationships between pressure at the head of the extruder and extruder output for (a) single-screw extruders and (b) intermeshing-screw twin-screw extruders for screws of given dimensions and polymers of given flow characteristics.

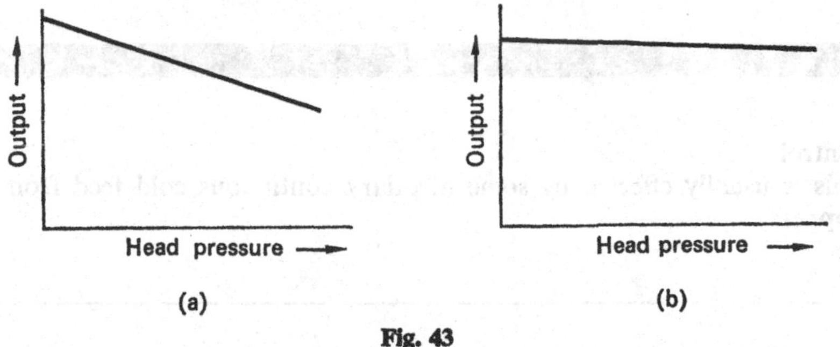

Fig. 43

High head pressures tend to occur with small orifice dies and in such circumstances give a lower output.

____-screw extruders give outputs which are only slightly sensitive to the area of the die orifice.

Twin

Frame 8.

It is possible to incorporate into the screw design, sections in which extra kneading and shearing is imposed on the polymer melt. Such extruders are sometimes used in the compounding (mixing) of polymer with additives.

Frame 9.

Give three main types of circumstances in which multi-screw extruders may be used.

(a) to overcome feeding problems with 'difficult' materials
(b) where it is desired to couple a high output with a high head pressure
(c) for compounding

Frame 10.

Since the pumping action is positive, overfeeding may cause the machine to overload and damage the bearings whilst underfeeding is obviously inefficient. It is therefore important to ________ the feed of material to the extruder.

---

control

This is usually effected by some auxiliary continuous cold feed from a hopper.

Frame 11.
When a multi-screw extruder is used it is difficult to design large thrust bearings at the rear of the screws to take up the head pressure. This can place restrictions on the output and screw speed of the machine.

Twin-screw machines also tend to be more expensive than single-screw machines of similar capacity.

Name three restricting factors which have limited the use of twin-screw extruders.

(a) the need for careful feed control
(b) restriction in output because of problems in thrust bearing design
(c) cost of the machine

Frame 12.
At the present time only a small proportion of extruders are of the twin-screw type. Their major fields of application are with unplasticised p.v.c. and for compounding purposes.

PART SEVEN

MISCELLANEOUS USES OF EXTRUDERS

Frame 1.

The screw pump extruder is often used in an intermediate stage in polymer processing.

The following are examples of its use:

(a) to receive material from polymerisation plant and extrude it into strands for subsequent granulation;
(b) to receive material in bulky form from mixing plant and extrude it into strands or sheet for subsequent granulation;
(c) to strain impurities from a polymer composition;
(d) as a mixing device;
(e) to extrude short lengths of tube known as parisons which are then inflated into bottles and other blow moulded articles;
(f) as part of an injection moulding machine.

These functions whilst technically very important are outside the scope of this booklet but when considered in conjunction with the use of the extruder in producing a wide variety of end products clearly show the crucial role of extrusion in plastics processing.

CRITERION TEST

Parts Five, Six and Seven

1. What is the most common way of determining the final external dimensions of extruded pipe?

2. What medium is used for cooling film made by the tubular process?

3. What problem occurs if tubular film is not sufficiently cooled before it is flattened out by the nip rollers.

4. Name the potential advantages of the cast-film process compared with the tubular-film process.

5. In the cast-film process how are the chill roll temperatures controlled?

6. In the cast-film process how is the draw down ratio controlled?

7. What is the first stage of the monofilament process?

8. For what reasons were multi-screw extruders first introduced?

9. Name two factors, other than cost, which have tended to limit the use of twin-screw extruders.

10. What name is given to short lengths of extruded tube which are subsequently inflated into bottles.

ANSWERS TO CRITERION TESTS

Part One

1. synthetic; organic; polymeric; non-elastomeric (at temperature of use)
2. fibres; surface coatings; adhesives
3. thermosetting; thermoplastics
4. an article made from a thermoplastics material will soften when heated; an article from a thermosetting plastic will remain set when heated
5. heat; pressure
6. pump; heating chamber; mixing device; pressure vessel

Part Two

1. feed; compression; melt (metering)
2. (a) to improve pumping efficiency in the feed zone
   (b) to reduce the effective channel depth in the melt zone and hence increase back pressure and produce a more uniform melt
3. a decreasing pitch screw
4. nylon
5. (a) to increase back pressure
   (b) to stop rotational flow of the melt
   (c) to hold back impurities
   (d) to hold back unplasticised materials
6. So that a decompression zone, from which volatiles may be removed, can be incorporated

Part Three

1. they become uncoiled or oriented
2. by increasing the melt temperatures or by increasing the draw down rate
3. by increasing the length of the die parallel
4. to ensure that the melt is fully consolidated
5. spiral mandrel die
6. because fish tail dies are very bulky, because high pressure are required to hold the die halves together and because the energy requirements to hold the large surface of such a die become economically prohibitive

Part Four

1. (a) entrapped air
   (b) monomer produced by polymer degradation in the extruder
2. polyethylene
3. places within the barrel or die head where polymer ceases to flow (stagnates); they should be avoided because such stagnant polymer decomposes with the decomposition products contaminating the rest of the polymer
4. false—they are almost invariably pseudoplastic
5. raise the melt temperature
6. powder feed

Parts Five, Six and Seven

1. by means of a sizing die
2. air
3. blocking
4. (a) lower haze
   (b) greater gloss
   (c) increased extrusion rate
5. by circulating water
6. adjusting the extrusion speed and the speed of the nip rolls
7. extrusion into a quench bath
8. a more positive acceptance of the feed material
9. (a) the need for careful feed control
   (b) problems of thrust bearing design
10. parisons